Science at Sea

Meeting Future Oceanographic Goals with a Robust Academic Research Fleet

Committee on Evolution of the National Oceanographic Research Fleet

Ocean Studies Board

Division on Earth and Life Studies

NATIONAL RESEARCH COUNCIL
OF THE NATIONAL ACADEMIES

THE NATIONAL ACADEMIES PRESS
Washington, D.C.
www.nap.edu

THE NATIONAL ACADEMIES PRESS 500 Fifth Street, N.W. Washington, DC 20001

NOTICE: The project that is the subject of this report was approved by the Governing Board of the National Research Council, whose members are drawn from the councils of the National Academy of Sciences, the National Academy of Engineering, and the Institute of Medicine. The members of the committee responsible for the report were chosen for their special competences and with regard for appropriate balance.

This study was supported by Contract/Grant No. N00014-05-G-0288, TO 17 between the National Academy of Sciences and the Office of Naval Research. Any opinions, findings, conclusions, or recommendations expressed in this publication are those of the author(s) and do not necessarily reflect the views of the organizations or agencies that provided support for the project.

International Standard Book Number-13: 978-0-309-14557-2
International Standard Book Number-10: 0-309-14557-0

Cover: Images from top to bottom counterclockwise: The image of original R/V *Atlantis* (1931-1964) was provided by Woods Hole Oceanographic Institution. The 1970 image of R/V *Melville* was provided by Scripps Institution of Oceanography, UC San Diego Libraries. The image of R/V *Alpha Helix* (University of Alaska Fairbanks) was provided by Bill Rook. The image of R/V *Pelican* was provided by Nancy Rabalais of the Louisiana Universities Marine Consortium. The image of R/V *Thompson* was provided by Kathleen K. Newell of the University of Washington. The image of R/V *Kilo Moana* was provided by SOEST, the University of Hawaii.

Additional copies of this report are available from the National Academies Press, 500 Fifth Street, N.W., Lockbox 285, Washington, DC 20055; (800) 624-6242 or (202) 334-3313 (in the Washington metropolitan area); Internet, http://www.nap.edu

Printed in the United States of America

THE NATIONAL ACADEMIES

Advisers to the Nation on Science, Engineering, and Medicine

The **National Academy of Sciences** is a private, nonprofit, self-perpetuating society of distinguished scholars engaged in scientific and engineering research, dedicated to the furtherance of science and technology and to their use for the general welfare. Upon the authority of the charter granted to it by the Congress in 1863, the Academy has a mandate that requires it to advise the federal government on scientific and technical matters. Dr. Ralph J. Cicerone is president of the National Academy of Sciences.

The **National Academy of Engineering** was established in 1964, under the charter of the National Academy of Sciences, as a parallel organization of outstanding engineers. It is autonomous in its administration and in the selection of its members, sharing with the National Academy of Sciences the responsibility for advising the federal government. The National Academy of Engineering also sponsors engineering programs aimed at meeting national needs, encourages education and research, and recognizes the superior achievements of engineers. Dr. Charles M. Vest is president of the National Academy of Engineering.

The **Institute of Medicine** was established in 1970 by the National Academy of Sciences to secure the services of eminent members of appropriate professions in the examination of policy matters pertaining to the health of the public. The Institute acts under the responsibility given to the National Academy of Sciences by its congressional charter to be an adviser to the federal government and, upon its own initiative, to identify issues of medical care, research, and education. Dr. Harvey V. Fineberg is president of the Institute of Medicine.

The **National Research Council** was organized by the National Academy of Sciences in 1916 to associate the broad community of science and technology with the Academy's purposes of furthering knowledge and advising the federal government. Functioning in accordance with general policies determined by the Academy, the Council has become the principal operating agency of both the National Academy of Sciences and the National Academy of Engineering in providing services to the government, the public, and the scientific and engineering communities. The Council is administered jointly by both Academies and the Institute of Medicine. Dr. Ralph J. Cicerone and Dr. Charles M. Vest are chair and vice chair, respectively, of the National Research Council.

www.national-academies.org

COMMITTEE ON THE EVOLUTION OF THE NATIONAL OCEANOGRAPHIC RESEARCH FLEET

RONALD KISS *(Co-chair)*, Webb Institute (retired), Rockville, Maryland
RICHARD PITTENGER *(Co-chair)*, Woods Hole Oceanographic Institution (retired), Massachusetts
FRANCISCO CHAVEZ, Monterey Bay Aquarium Research Institute, Moss Landing, California
MARGO EDWARDS, University of Hawaii, Manoa
RANA FINE, University of Miami, Florida
NANCY RABALAIS, Louisiana Universities Marine Consortium, Chauvin
ERIC SALTZMAN, University of California, Irvine
JAMES SWIFT, University of California, San Diego
WILLIAM WILCOCK, University of Washington, Seattle
DANA YOERGER, Woods Hole Oceanographic Institution, Massachusetts

Staff

DEBORAH GLICKSON, Associate Program Officer
JEREMY JUSTICE, Program Assistant

Preface

The vastness of the ocean invites, but then defies, simple description. The ocean encompasses more than 70 percent of the Earth's surface, with depths of almost 11,000 meters. The ocean system plays an essential role in weather and climate. Winds drive the continual circulation of the ocean currents. Some parts of the ocean are ice-covered all or part of the year. The ocean has a very large heat capacity, is a major carbon dioxide sink, and has the ability to buffer, absorb, and disperse chemicals. Billions of people are fed with biomass from the ocean, and the oceans are important avenues for commerce, recreation, and national defense. The ocean preserves a record of Earth's climatic processes and an archaeological record of human civilization. Although the ocean is large, it is not immune to natural or human-induced change. For example, the ocean is warming and acidifying, and the world's fisheries are severely stressed. Marine debris from both ships and land is cluttering the ocean, while nutrient pollution and toxic runoff pose threats to marine life and human health.

The ocean, vital though it may be, is extraordinarily difficult to sense and model. The endlessly complex and variable seas are undersampled. Oceanographic research is still in discovery mode, with each year bringing unimagined new surprises. Studying the biota or the shape of the ocean floor requires sensing or traversing through thousands of meters of pitch black, frigid water at enormous pressures. These conditions define the scientific challenge we call oceanography.

For centuries, ships have provided the primary means of observing and measuring ocean parameters. Technology and invention have produced many improvements, with moored and hard-wired sensors and with an increasingly sophisticated family of autonomous vehicles. The Committee on Evolution of the National Oceanographic Research Fleet was convened by the National Research Council to assist the Office of Naval Research and the National Science Foundation in determining how rapid advancements in ocean observing technology and rising costs will impact the future U.S. academic research fleet relative to Navy needs.

An excellent group of scientists with expertise in physical, chemical, and biological oceanography, marine geology and geophysics, atmospheric science, ocean engineering, naval architecture, and ship operations and policy volunteered their time and talent for this study. The committee met four times over the course of six months in 2008 and 2009. In open sessions in Washington, D.C., and Woods Hole, Massachusetts, the committee called upon a cadre of marine experts to shed light on its charge. The primary goal of these meetings was to understand how developments in both science and technology would impact oceanographers' needs for research vessels.

RADM Richard Pittenger, *co-chair*
Ronald K. Kiss, *co-chair*

Acknowledgments

This report was greatly enhanced by the participants of the meetings held as part of this study. The committee would first like to acknowledge the efforts of those who gave presentations at meetings: Vernon Asper (University of Southern Mississippi), Brian Calder (University of New Hampshire), David Fisichella (Woods Hole Oceanographic Institution), Matt Hawkins (National Science Foundation), Dave Hebert (University of Rhode Island), Frank Herr (Office of Naval Research), Stanford Hooker (National Aeronautics and Space Administration), Tajr Hull (National Oceanic and Atmospheric Administration), Tim Killeen (National Science Foundation), Dennis McGillicuddy (Woods Hole Oceanographic Institution), Mike Prince (Moss Landing Marine Laboratories), Dan Rolland (Alion Science), Suzanne Skelley (National Oceanic and Atmospheric Administration), Robin Williams (Robin A. Williams and Co. Ltd.), and Marc Willis (Oregon State University). These talks helped set the stage for fruitful discussions in the closed sessions that followed.

The committee is also grateful to a number of people who provided important discussion and/or material for this report: Steven Ackleson, Annette DeSilva, Rose Dufour, Bauke Houtman, Mike Purcell, Elizabeth Rios Brenner, Tim Schnoor, John Toole, and Bob Weller.

This report has been reviewed in draft form by individuals chosen for their diverse perspectives and technical expertise, in accordance with procedures approved by the National Research Council (NRC's) Report Review Committee. The purpose of this independent review is to provide candid and critical comments that will assist the institution in making its published report as sound as possible and to ensure that the report meets institutional standards for objectivity, evidence, and responsiveness to the study charge. The review comments and draft manuscript remain confidential to protect the integrity of the deliberative process. We wish to thank the following individuals for their participation in the review of this report:

James Bellingham, Monterey Bay Aquarium Research Institute, Moss Landing, California
DelWayne Bohnenstiehl, North Carolina State University, Raleigh
Robert Embley, National Oceanic and Atmospheric Administration, Newport, Oregon
James McWilliams, University of California, Los Angeles, California
Christopher Measures, University of Hawaii, Manoa, Honolulu
Mike Prince, Moss Landing Marine Laboratories, California
Martin Visbeck, University of Kiel, Germany
Dick West, Consortium for Oceanographic Research and Education (retired), Coventry, Rhode Island
Peter Wiebe, Woods Hole Oceanographic Institution, Woods Hole, Massachusetts
Robin Williams, Robin A. Williams & Co. Ltd., Barry, Wales, United Kingdom
Marc Willis, Oregon State University, Corvallis

Although the reviewers listed above have provided many constructive comments and suggestions, they were not asked to endorse the conclusions or recommendations nor did they see the final draft of the report before its release. The review of this report was overseen by **Robert A. Duce**, Texas A&M University, College Station. Appointed by the National Research Council, he was responsible for making certain that an independent examination of this report was carried out in accordance with institutional procedures and that all review comments were carefully considered. Responsibility for the final content of this report rests entirely with the authoring committee and the institution.

Contents

Summary

Oceanography has traditionally been an observational science, with researchers dependent on ships to provide them with access to the oceans. Since 1971, the U.S. academic research fleet has been managed through the University-National Oceanographic Laboratory System (UNOLS), a consortium that unites research institutions, federal agencies, and state and private interests.

Requiring advice on scientific and technological issues that may affect the evolution of the U.S. academic research fleet, the Office of Naval Research (ONR) and the National Science Foundation (NSF) asked the National Academies to provide near-term advice on how rapid advancements in ocean observing technology and rising costs will impact the future fleet relative to Navy needs. The Academies convened the Committee on Evolution of the National Oceanographic Research Fleet to examine a number of factors including the impacts of advanced technologies such as autonomous vehicles and ocean observing systems on data collection; the most important factors in research vessel design; the impacts of evolving modeling and remote sensing approaches on research operations; the impact of rising costs of research vessel operations on the ability to conduct oceanographic research in the future; and the usefulness of partnering mechanisms, such as UNOLS, to support national oceanographic research objectives.

FUTURE SCIENCE NEEDS

Societal awareness of the ocean's critical role in complex environmental and natural hazards issues has increased dramatically over the past few years. As a result, many topics previously of interest only to a select group of oceanographers (including ocean acidification, carbon and biogeochemical cycling, ocean circulation, ocean-atmosphere fluxes, harmful algal blooms, undersea volcanic eruptions, and tsunami generation) are now being viewed as essential for national and worldwide health and security.

These issues will require a fundamental understanding of complex and interwoven processes, grounded in sustained ocean observations. The future ocean research agenda will be driven by diverse disciplinary and interdisciplinary research across a broad range of spatial and temporal observational scales. *Key to the study of these issues is the U.S. academic research fleet, which provides an essential, enabling resource for the nation. Scientific demands on the U.S. academic fleet are likely to increase in future years. However, aging ships and evolving technology require fleet modernization and recapitalization to maintain the nation's leadership in ocean research.*

> **Recommendation: Federal agencies supporting oceanographic research should implement one comprehensive, long-term research fleet renewal plan to retain access to the sea and maintain the nation's leadership in addressing scientific and societal needs.**

The paradigm of a single investigator going to sea to examine a specific research problem has given way to larger scientific teams engaged in multidisciplinary research cruises to study more complex questions. Technological developments in autonomous mobile platforms, fixed observatories, sensors, remote sensing, and modeling will continue to increase scientific understanding of the ocean environment but will not obviate the need for research vessels. *The fleet of the future will be required to support increasingly complex, multidisciplinary, multi-investigator research projects, including those in support of autonomous technologies, ocean observing systems, process studies, remote sensing, and modeling.* Adaptable, technologically advanced Global class vessels will be needed. Critical interdisciplinary research on coastal margins will require capable Regional class vessels that operate in shallower depths.

TECHNOLOGICAL ADVANCEMENTS

The growing use of autonomous vehicles has already changed the role of the research fleet. Ships are increasingly used as platforms to support synchronous operations of multiple vehicles, requiring the ability to carry more instruments, equipment, and personnel. Ocean observing systems will also increase pressure on the academic research fleet. Ships will be needed to support installation, operations, and maintenance of observatory infrastructure, as well as sensor package deployment and novel science programs. *Ocean observatories and autonomous vehicles will impact future vessel design requirements for acoustic communications, deck space, payload, berthing, launch and recovery, and stability. Servicing ocean observatories and launching and recovering autonomous vehicles will result in increased demands for ship time.*

Satellite data and more advanced ocean modeling are providing scientists with valuable analysis tools to place their observations of ocean variability in context across a spectrum of spatial and temporal scales. Increased access to satellite remote sensing data and broader ship-to-shore communications bandwidth will allow for interdisciplinary process studies that integrate real-time imagery. This will strengthen the need for ship-based calibration and validation of satellite data and will increase the pressure for robust ship-to-shore satellite communications. This technology will also provide greater opportunity for land-based researchers to participate remotely in research cruises, increasing the efficiency of ship-based science. T*here is a need for increased ship-to-shore bandwidth, in order to facilitate real-time, shore-based modeling and data analysis in support of underway programs, allow more participation of shore-based scientists, and increase opportunities for outreach.*

VESSEL DESIGN

Future oceanographic vessels will continue to support widely diverse research objectives, with increased pressure to facilitate multidisciplinary, multi-investigator research. *Supporting future research needs will require both highly adaptable general purpose ships and specialized vessels. Some vessels should be capable of operating in high latitudes and high sea states. More capable Coastal, Regional, and Global class ships will also be needed.* Larger science parties and more complex technology will require more laboratories, deck space, and accommodations. Trends toward increasing beam, length, draft, and displacement and the economy of scale present in larger hulls suggest that investments should be made in larger, more capable vessels in any size class.

Some existing Navy-built research vessels have suffered from poorly defined performance specifications, leading to less-than-optimal research vessels. The current Navy ship acquisition process does not emphasize inclusion of the scientific community in decision making regarding design and specifications. The process led by NSF in its design and procurement of the Alaska Region Research Vessel (ARRV) is a refined continuation of efforts to include ocean researchers in ship design and construction. *Development of the NSF-sponsored ARRV has benefited from community-driven ship design, allowing users to participate more fully and to create optimal designs within cost constraints.*

Recommendation: All future UNOLS ship acquisitions, beginning with the planned Ocean class vessels, should involve the scientific user community from the preconstruction phase through post-delivery of the ship.

SHIP TIME COSTS

Total operating costs for the UNOLS fleet increased 75 percent from 2000 to 2008, driven mainly by crew and fuel cost factors. Recent market volatility of crude oil led to extremely high fuel costs in 2008 and more expensive daily ship rates. Over the same period, the total number of operating days decreased by 13 percent. The continued push for operating efficiency may lead to longer lead times for research projects and reductions in the ability of the future fleet to accommodate late-breaking scientific and funding opportunities.

The increasing cost of ship time and the economies of scale associated with larger ships may lead to greater use of the Global class vessels, which have laboratories, deck space, and berthing capabilities that can support multiple science operations. With these vessels, complex programs are less likely to require multiple legs, thus lowering operational costs.

Recommendation: The future academic research fleet requires investment in larger, more capable, general purpose Global and Regional class ships to support multidisciplinary, multi-investigator research and advances in ocean technology.

PARTNERSHIPS

The UNOLS partnership between federal agencies, academic institutions, and state and private interests successfully serves national oceanographic research objectives and is anticipated to continue in the face of changing science priorities and technological advances. *The UNOLS consortium management structure is sound and is of benefit to research institutions, federal agencies, and state and private interests. The federal agency partnerships that capitalize and support the academic research fleet, particularly between the Navy and NSF, have a proven record of cost savings and asset sharing. However, there are many assets that are not integrated with UNOLS, leading to suboptimal use of the full U.S. research fleet.* This leads to a mismatch between available ship time and research needs to support national goals, a trend that is likely to continue in the future.

In particular, opportunities exist to better integrate icebreakers operated by the U.S. Coast Guard and supported by NSF's Office of Polar Programs with the UNOLS management structure, and to fulfill some part of the National Oceanic and Atmospheric Administration (NOAA's) identified needs for significantly more ship time by utilizing UNOLS unfunded ship days. A stronger partnership between UNOLS and NOAA would allow NOAA to better fulfill its mission and UNOLS to increase efficient use of the fleet.

Recommendation: NOAA should identify which of its 13,200 unmet annual ship day needs could be supported by the UNOLS fleet. NOAA and UNOLS should work together to develop a long-term plan to increase the usage of UNOLS ships in support of the NOAA mission.

Recommendation: The NSF Division of Ocean Sciences, the NSF Office of Polar Programs, and the U.S. Coast Guard should improve coordination of ship operations and support between the UNOLS and polar research fleets.

1

The U.S. Academic Research Fleet

The academic research fleet provides U.S. and international users with access to the ocean—from the nearshore coastal zones to deep, remote regions far from land. Research vessels provide oceanographers with opportunities to study issues of increasing societal relevance, including the ocean's role in climate, natural hazards, economic resources, human health, and ecosystem sustainability. A highly capable fleet of ships also provides a platform for innovative basic research in chemical, biological, and physical oceanography; marine geology and geophysics; atmospheric science; and emerging interdisciplinary areas.

Reports from the U.S. Commission on Ocean Policy (USCOP) and the Joint Subcommittee on Ocean Science and Technology (JSOST) have recognized the academic fleet as an essential component of ocean research infrastructure (U.S. Commission on Ocean Policy, 2004; Joint Subcommittee on Ocean Science and Technology, 2007). At the same time, there is community concern that the fleet is in dire need of both modernization and recapitalization (i.e., U.S. Commission on Ocean Policy, 2004; Malakoff, 2008; UNOLS Fleet Improvement Committee, 2009).

BACKGROUND

The UNOLS Consortium

The U.S. academic research fleet is managed through the University-National Oceanographic Laboratory System (UNOLS; Box 1-1), a consortium that unites research institutions, federal agencies, and state and

BOX 1-1
What Is UNOLS?

- The UNOLS mission is to "provide a primary forum through which the ocean research and education community, research facility operators and the supporting federal agencies work cooperatively to improve access, scheduling, operation, and capabilities of current and future academic oceanographic facilities."
- 18 UNOLS institutions operate shared-use facilities, including 22 research vessels, a National Deep Submergence Facility, a National Oceanographic Aircraft Facility, and a National Oceanographic Seismic Facility.
- UNOLS acts in an advisory role to facility operators and to supporting federal agencies, but it is not itself a funding agency or a facility operator.
- UNOLS supports community-wide efforts to provide broad access to oceanographic research facilities; continuous improvement of existing facilities; and planning for future oceanographic facilities.

SOURCE: UNOLS website (*www.unols.org*) and UNOLS Fleet Improvement Committee (2009).

private interests. Although the academic fleet has existed since before World War II (history provided in Appendix A), the UNOLS management structure was not established until 1971, based on a recommendation of the Stratton Commission report *Our Nation and the Sea* (Commission on Marine Science, 1969; Byrne and Dinsmore, 2000; Bash, 2001). From 18 original operating institutions (Byrne and Dinsmore, 2000), by 2009 membership had grown to 61 institutions representing 26 states and Panama, Puerto Rico, and Bermuda (Appendix B). UNOLS coordinates the schedules of 22 vessels berthed in 13 states and Bermuda.

UNOLS assists federal and states agencies in performing their seagoing responsibilities. The National Science Foundation (NSF), Office of Naval Research (ONR), National Oceanic and Atmospheric Administration (NOAA), U.S. Geological Survey (USGS), Minerals Management Service (MMS), and U.S. Coast Guard (USCG) support the UNOLS consortium through a cooperative agreement. Other agencies, including the Environmental Protection Agency (EPA), National Aeronautics and Space Administration (NASA), U.S. Army Corps of Engineers (USACE), and Department of Energy (DOE) support ship time on UNOLS vessels (Annette DeSilva, personal communication, 2009). State funds and private resources are also used to support the academic fleet.

The UNOLS Fleet

The current UNOLS fleet (Table 1-1) consists of six classes of ships (Federal Oceanographic Facilities Committee, 2001; Interagency Working Group on Facilities, 2007; UNOLS Fleet Improvement Committee, 2009). Of these, the Global, Ocean, Intermediate, and Regional classes have been most likely to be built or acquired with federal funds (Interagency Working Group on Facilities, 2007).

Global class vessels are large, high-endurance ships capable of working worldwide. They are able to stay at sea for 50 or more days and can carry 30-38 scientists. Two of the six Global class ships are specialized: *Atlantis* is the tender for the *Alvin* deep submersible, and *Marcus Langseth* is a seismic ship. While the four other Global class vessels are general purpose, each also carries specialized equipment (e.g., long coring ability on *Knorr*). Intermediate class ships are medium-endurance, ocean-ranging vessels with berths for 18-20 scientists. Three of the five Intermediate vessels (*Endeavor*, *Oceanus*, and *Wecoma*) are approaching the end of their service lives and are projected to retire in 2010. The Ocean class was envisioned in the 2001 Federal Oceanographic Facilities Committee (FOFC) report *Charting the Future for the National Academic Research Fleet* as a replacement for the aging, less capable Intermediate class (Federal Oceanographic Facilities Committee, 2001). These general purpose, oceangoing vessels are designed to have ranges up to 40 days and accommodations for 25 scientists. There is currently one Ocean class vessel, *Kilo Moana*, with three more planned. Regional and Regional/Coastal class vessels serve coastal oceanography needs, with 30-day endurance and capacity for up to 20 scientists. There are two main distinctions between these classes: all four of the Regional/Coastal vessels were funded through state sources, while two of the three Regional ships were acquired by NSF; and Regional class ships generally have a little more range and endurance than Regional/Coastal vessels, which would work closer to the coast and often conduct shorter cruises closer to port. Local class ships work in the nearshore environment, with an endurance of about 20 days and berthing for 15 or fewer scientists. Most Local class vessels are owned by individual institutions.

The Navy has historically been a strong supporter of academic ocean research in the United States. In addition to funding scientific research and instrument development, there is a long and well-invested portfolio of assets in the U.S. academic research fleet (see Appendix A and Table 1-1, respectively, for past and current Navy-funded UNOLS ships). The Navy currently owns five of the six Global vessels and the sole Ocean class vessel in the UNOLS fleet, and has traditionally capitalized the largest ships of the UNOLS fleet. NSF owns the Global class *Marcus Langseth*, three Intermediate class vessels, and several smaller ships. NSF funds

Table 1-1 The 2009 UNOLS Research Fleet (adapted from *www.unols.org*; used with permission from UNOLS)

Operating Institution	Ship	Year Built/ Converted	Owner	Length (ft)
Global				
Scripps Institution of Oceanography (SIO)	*Melville*	1969	Navy	279
Woods Hole Oceanographic Institution	*Knorr*	1970	Navy	279
University of Washington	*Thomas G. Thompson*	1991	Navy	274
Scripps Institution of Oceanography	*Roger Revelle*	1996	Navy	274
Woods Hole Oceanographic Institution	*Atlantis*	1997	Navy	274
Lamont-Doherty Earth Observatory	*Marcus Langseth*	2008	NSF	235
Ocean				
University of Hawaii	*Kilo Moana*	2002	Navy	186
Intermediate				
Harbor Branch Oceanographic Institute, Florida Atlantic University (FAU)	*Seward Johnson*	1985	FAU	204
Oregon State University	*Wecoma*	1976	NSF	185
University of Rhode Island	*Endeavor*	1977	NSF	185
Woods Hole Oceanographic Institution	*Oceanus*	1976	NSF	177
Scripps Institution of Oceanography	*New Horizon*	1978	SIO	170
Regional				
Bermuda Institute for Ocean Sciences (BIOS)	*Atlantic Explorer*	2006	BIOS	168
Duke University/University of North Carolina	*Cape Hatteras*	1981	NSF	135
Moss Landing Marine Laboratories	*Point Sur*	1981	NSF	135
Regional/Coastal				
University of Delaware (UD)	*Hugh R. Sharp*	2005	UD	146
Scripps Institution of Oceanography	*Robert Gordon Sproul*	1981	SIO	125
Louisiana Universities Marine Consortium (LUMCON)	*Pelican*	1985	LUMCON	116
University of Miami (UM)	*Walton Smith*	2000	UM	96

Table 1-1 Continued

Operating Institution	Ship	Year Built/ Converted	Owner	Length (ft)
Local				
University System of Georgia/ Skidaway (UG/SKIO)	*Savannah*	2001	UG/SKIO	92
University of Minnesota, Duluth (UMD)	*Blue Heron*	1985	UMD	86
University of Washington	*Clifford Barnes*	1966	NSF	66

the majority of ship operating days (58 percent between 2000 and 2009; Annette DeSilva, personal communication, 2008) and fleet operating costs (63 percent in 2007) (UNOLS Fleet Improvement Committee, 2009). By comparison, the Navy utilized an average of 17 percent of UNOLS ship operating days in the same time period.

REPORT SCOPE

The Navy has committed to build two new Ocean class vessels, scheduled to enter service in 2014 and 2015, with ONR as the mission sponsor. Both ONR and NSF are interested in the impact of evolving science needs, rapid technological advancements, and increasing operational costs on future research fleet capabilities. They have asked the National Academies to carry out an independent and objective assessment of the scientific and technological issues that may affect the evolution of the UNOLS fleet (see Box 1-2 for Statement of Task).

Because of the long lifespan of the research fleet assets (often 30 or more years), there is a strong emphasis on adequate planning in the present to make sure the fleet remains capable of supporting future scientific research. This report investigates future vessel needs, including fleet mix, but does not address or recommend a specific number of ships needed. In the same vein, an "optimal mix" of autonomous and remote platforms, observing systems, and remote sensing is not addressed because of an inability to predict future disruptive technologies that could revolutionize the field of oceanography. This report is also not intended to impact the major design elements of the two planned Ocean class vessels, which were in development when the study was commissioned.

Primary technology drivers for this study include recent investments in ocean observing systems (e.g., NSF's Ocean Observing Initiative [OOI] and NOAA's investment in the Integrated Ocean Observing System [IOOS]) and associated long-duration sensor packages; growth in the use

Box 1-2
Statement of Task

In support of the need for oceanographic fleet replacement, ONR is currently in the early design process for the first of two new Ocean class ships and requires near-term advice on how the rapid advancements in ocean observing technology and the impacts of rising costs will impact the future fleet relative to Navy needs. Therefore, ONR and NSF have requested that the National Research Council (NRC) appoint an ad hoc committee to review the scientific and technological issues that may affect the evolution of the UNOLS academic fleet, including:

1. How technological advances such as autonomous underwater vehicles and ocean observing systems will affect the role and characteristics of the future UNOLS fleet with regard to accomplishing national oceanographic data collection objectives.
2. The most important factors in oceanographic research vessel design. Does specialized research needs dominate the design criteria and, if so, what are the impacts on costs and overall availability?
3. How evolving modeling and remote sensing technologies will impact the balance between various research operations such as ground-truthing, hypothesis testing, exploration, and observation.
4. How the increasing cost of ship time will affect the types of science done aboard ships.
5. The usefulness of partnering mechanisms such as UNOLS to support national oceanographic research objectives.

and maturity of remotely operated and autonomous vehicles; and increasingly sophisticated modeling and remote sensing.

Evolving directions in scientific research and their expected impacts on research vessel design are also examined in the context of past experiences and present trends. The fleet is required to support a broad range of oceanographic missions, including those in physical, biological, and chemical oceanography; marine geology and geophysics; and atmospheric science. For this reason, ONR's intent with its Global and Ocean class vessels has been to provide a general purpose platform for science (Frank Herr, personal communication, 2009). The committee has identified design requirements dictated by research needs, with a discussion of the costs entailed.

Capital and life-cycle costs are also strong drivers of the academic fleet. Construction costs are dependent on shipyard labor needs and the cost of raw materials such as steel. Crew salaries and benefits costs have historically been the largest percentage of vessel operating costs, although rising fuel prices from 2005 to 2008 contributed to increasingly higher overall operating costs (UNOLS Fleet Improvement Committee, 2009).

STUDY APPROACH AND INFORMATION NEEDS

To properly evaluate the factors and demands that may drive future fleet needs, the committee considered a number of issues. Major trends in future oceanographic research were examined as a necessary complement to technological advances. The committee studied many recent community planning documents and agency strategic plans for future ocean science directions to evaluate these needs. During the information gathering process, presentations by and discussions with representatives of federal agencies, scientists, engineers, shipboard scientific support personnel, and ship operators were used to discern trends in science usage, new technologies, and vessel needs. The committee chose not to explore quantitative analyses of recent publications or conference abstracts, because members did not feel that such analyses would provide accurate, forward-looking measures of community scientific trends or changing fleet needs. Statistics related to fleet operating costs and usage trends were obtained from the UNOLS Office and examined by the committee. Due to the minor differences between their respective classes, Regional and Regional/Coastal vessels were considered together and are often used interchangeably in this report.

The academic research fleet has been studied often. Federal advisory boards, interagency groups, and the UNOLS Fleet Improvement Committee have all expended considerable effort discussing the status of the fleet, projections into the future, and renewal plans. These prior reports are summarized below.

Past Assessments

In 1999, *The Academic Research Fleet* was written in response to a request from NSF's Science Advisory Board (Fleet Review Committee, 1999). The committee was asked to evaluate current and future vessel requirements for NSF oceanographic research and to report on the overall management structure for the research fleet. Among its findings and recommendations, the report recognized that the strength of the UNOLS system was in the highly trained crew and ship operators that supported seagoing research. UNOLS management and practices were also commended. The report indicated some concern about a potential decreasing trend in fleet use and called for the introduction of new technologies into the fleet and improvements in training and quality control. The report recommended that federal agencies prepare and coordinate long-range plans for the academic fleet.

An NSF-sponsored workshop held in 2000, *Assessment of Future Science Needs in the Context of the Academic Oceanographic Fleet*, examined fleet needs in the context of future science research and new observational technology. Workshop participants concluded that new observational tools and systems would not reduce or replace the need for an academic research fleet. Instead, future research and tools would increase demand for ship time and for more capable ships (Cowles and Atkinson, 2000).

NSF's 2001 report *Ocean Sciences at the New Millennium* asserted that "maintaining a modern, well-equipped research fleet is the most basic requirement for a healthy and vigorous research program in the ocean sciences" and strongly recommended that a long-term plan for fleet renewal be enacted (National Science Foundation, 2001).

That same year, FOFC, a federal interagency committee of the National Oceanographic Partnership Program (NOPP), released *Charting the Future for the National Academic Research Fleet* (Federal Oceanographic Facilities Committee, 2001). That report responded to data in *The Academic Research Fleet* by setting forth a renewal strategy for the academic research fleet, with the underlying assumption that fleet capacity[1] would be maintained while capabilities were increased. It outlined a 20-year plan for adding 10-13 additional vessels to the academic fleet, discussed planning for technology upgrades and updating ship concept designs and science mission requirements, and proposed the introduction of Ocean class vessels as replacements for aging and less capable Intermediate vessels of the fleet. The plan was to be revised at least once every 5 years to account for changing science needs.

In its 2004 report *An Ocean Blueprint for the 21st Century*, the U.S. Commission on Ocean Policy praised the UNOLS fleet renewal plan outlined in *Charting the Future for the National Academic Research Fleet*. However, the members of the commission expressed concern that at the time of their report there had been no move to implement the plan or provide funding for fleet renewal (U.S. Commission on Ocean Policy, 2004).

In 2007, the Interagency Working Group on Facilities (IWGF), a successor to FOFC established by JSOST, released the *Federal Oceanographic Fleet Status Report* (Interagency Working Group on Facilities, 2007). The IWGF report described the current status and planned renewal activities of federally-owned academic ships more than 40 meters in length and other federal fleet assets in the 2007-2015 time frame. Renewal plans put forth in the 2001 FOFC report either were not addressed in this report or

[1]Fleet capacity was defined as 3,600 days, equal to the total operational days averaged over the previous 5-year interval (1997 to 2001).

were scaled down, with the exception of a replacement for the seismic vessel *Maurice Ewing*.

The most recent assessment of the fleet was done in 2009 by the UNOLS Fleet Improvement Committee (2009). Its *Fleet Improvement Plan* addressed the needs of the U.S. research fleet through 2025. The report recommended that it was necessary for the academic research fleet to increase beyond the levels projected in the *Federal Oceanographic Fleet Status Report* and that federal agencies should proceed with existing and planned fleet renewal activities. It was noted that the current planned renewal contains fewer ships than was recommended in the 2001 *Charting the Future for the National Academic Research Fleet* plan.

ORGANIZATION OF THIS REPORT

This report addresses oceanographic research and technology needs that should influence the development of the U.S. academic fleet in the next 10-20 years. Chapter 2 surveys future science trends that will impact fleet usage in the near future, while Chapter 3 provides a discussion on specific technological advancements that may impact research vessel needs. Chapters 2 and 3 both address aspects of the first and third components of the Statement of Task (Box 1-2; Tasks 1 and 3). Research vessel design factors and criteria (Task 2) are outlined in Chapter 4, while fleet costs and the resulting impact on research (Task 4) are discussed in Chapter 5. Chapter 6 discusses the UNOLS partnership structure and its usefulness (Task 5). A summary and recommendations are included in Chapter 7. Relevant items from the Statement of Task are listed at the beginning of each chapter.

2

Future Science Needs

How technological advances such as autonomous underwater vehicles and ocean observing systems will affect the role and characteristics of the future UNOL fleet with regard to accomplishing national oceanographic data collection objectives.

How evolving modeling and remote sensing technologies will impact the balance between various research operations such as ground-truthing, hypothesis testing, exploration, and observation.

The future ocean research agenda will address major societally relevant issues, including the ocean's role in climate change, ecosystem health and sustainability, marine impacts on human health, management and exploitation of natural resources, and improving the predictability of natural hazards and maritime safety. These are inherently multidisciplinary challenges, involving the physical, chemical, biological, and geological sciences and allied fields such as air-sea interaction and atmospheric science. As an example, understanding the role of the oceans in the Earth's climate system involves assessing the influence of topography on ocean circulation, storage and redistribution of heat, salt and carbon dioxide (CO_2) in the ocean, exchange of energy between the ocean and atmosphere, biogeochemical changes influencing ocean uptake and release of greenhouse gases, and the impacts of climate change on marine ecosystems. In addition, basic and exploratory oceanographic research will continue to be needed.

New technologies (e.g., autonomous mobile systems, fixed seafloor observatories, and remote sensing and modeling) have revolutionized traditional observation-limited oceanographic research, drastically increasing both the amount of data collected and the sophistication of analysis and assimilation. This does not lessen the continuing need for a versatile, technologically capable fleet of research vessels to support oceanographic research. Complex chemical and biological measurements will continue to require shipboard laboratories, and advanced technologies still require ships as platforms and tenders. Technological advances are discussed in detail in Chapter 3 but are introduced here in the context of major science research drivers.

This chapter provides a brief survey of major research trends and needs that will influence the use and design of the future academic fleet. It is not intended as a comprehensive inventory of future oceanographic directions, which can be found in recent community planning documents and agency strategic plans (i.e., Baker and McNutt, 1996; Young et al., 1997; Trenberth and Clarke, 1998; Jumars and Hay, 1999; National Science Foundation, 2001; Ridge 2000 Program, 2001; Liss et al., 2004; MARGINS Office, 2004; MESH Workshop Steering Committee, 2005; National Oceanic and Atmospheric Administration, 2005; Daly et al., 2006; Joint Subcommittee on Ocean Science and Technology, 2007; National Oceanic and Atmospheric Administration, 2008). For organizational ease, this chapter is broken down by disciplinary needs, with the recognition that there is considerable overlap due to the multidisciplinary nature of major scientific questions driving oceanographic research. Several case studies are shown in boxes to help to illustrate multidisciplinary oceanographic research programs that will incorporate new technology and drive the need for adaptable, capable research vessels. Box 2-1 is an example of a current research problem; Boxes 2-2 through 2-5 are hypothetical, near-future scenarios.

PHYSICAL OCEANOGRAPHY

Physical oceanography research focuses on the physical properties and dynamics of ocean processes. Future research needs are directed toward the role of ocean circulation and properties in climate change and the global carbon cycle. Global arrays of autonomous platforms and sensors and ship-based hydrography and process studies are essential to progress in these research needs. Ocean circulation changes in the full water column have been linked to a wide range of climatic variations that are of clear and critical interest to society. Ship-based measurements are needed (Hood et al., 2009) to

- Reduce uncertainties in global freshwater, heat, and sea level budgets;
- Determine ocean ventilation and circulation pathways and rates using chemical tracers; and
- Determine the variability in and controls of water mass formation and properties.

Physical oceanography has been transformed by the numerous autonomous sampling devices currently available (e.g., moorings, drifters, floats, autonomous underwater vehicles [AUVs], gliders), which increased sampling in the upper ocean to a rate and density unparalleled by ships. Research vessels are still needed to measure large-scale changes in ocean heat and freshwater fluxes, deep ocean variability below 2000 meters,[1] and the anthropogenic inventory of CO_2 (Garzoli et al., 2009). Many climatically important carbon and related transient tracer parameters cannot be measured from autonomous devices with present-day technology, and few floats, gliders, and AUVs are able to operate to the full depth of the water column. While some of these instruments will operate to greater ocean depths in the future, there will continue to be parts of the deep ocean that cannot be reached without ship-based equipment. High-quality, ship-based observations will also continue to be essential for calibration of water column measurements made from autonomous devices.

The deep ocean accounts for more than half of the total natural oceanic carbon inventory. As anthropogenic carbon begins to invade the deep oceans in nonhomogeneous ways, it will continue to be critical to monitor changes in deep ocean carbon content. For example, observations of transient tracers (Willey et al., 2004), particularly in the high latitudes, strongly suggest that ventilation by atmospheric gases is more rapid than previously estimated. In addition, observations of biogeochemical parameters show greater-than-expected variations at depth, which suggest that natural and/or climate-induced change is having a greater effect on deep waters than was previously assumed.

Ship-based information and global sensor arrays will be critical to evaluate ocean general circulation models and provide data constraints for inverse models. Process studies, such as deliberate tracer and mixing experiments (see Box 2-1), will continue to require research vessels as platforms for science, as will the continuation of established time series (Bermuda Atlantic Time Series, Hawaii Ocean Time Series, Atlantic Meridional Overturning Circulation at 26.5°N, Line W mooring and hydrographic time series in the North Atlantic). Global surveys are the most effective

---

[1] This lower volume between the seabed and 2000 meters depth, taken over its total global area, constitutes more than 50 percent of total oceanic volume.

BOX 2-1
Salt Fingers Show Vigorous Ocean Mixing

Incomplete understanding of ocean mixing has been a limitation in predicting Earth's future climate, specifically for modeling the oceans' absorption and storage of climatically important properties, including heat and carbon dioxide. For decades, scientists have known that in certain parts of the ocean, layers of warm salty water of subtropical origin overlay cooler and fresher water of Antarctic origin. The interaction of these layers creates "salt fingers," salty staircases driven by small-scale convection. Using data from a ship-based process study in the North Atlantic, it was discovered that salt fingers transform the temperature and salinity structure of water entering the Caribbean Sea. The resultant increase in salinity and density preconditions the Antarctic water at intermediate depths for sinking at high latitudes of the North Atlantic. Sampling from a ship allowed direct measurement of the vertical spreading rate of a passive tracer injected in the middle of a staircase. This allowed quantification of the effect of the salt fingering on enhanced mixing within the thermocline. The resultant mixing is an order-of-magnitude greater than the background mixing due to the breaking of internal waves (Schmitt et al., 2005). These results highlight the need to include mixing due to salt fingering in climate models.

means for quantifying the variability of a large suite of physical and biogeochemical parameters, and global full-depth reassessments of the temperature, salinity, carbon, and related tracer distributions are a critical component of a global ocean and climate observing system (Hood et al., 2009). These surveys will continue to require Global class vessels, which are the only U.S. ships with sufficient endurance and range. Large-scale observing networks such as the Argo array, composed of 3000 profiling floats equipped with conductivity-temperature-depth (CTD) sensors, revolutionized scientists' view of the ocean by providing extraordinary temporal and spatial coverage in the upper 2000 meters of the water column. However, these arrays can increase pressure for research vessels for deployment and calibration via ship-based datasets.

Ocean acoustics, the branch of physical oceanography that studies the physics of sound in the ocean, will also continue to require ship-based experimentation and advances in acoustically quiet technologies. Shallow-water, high-frequency, and long-range acoustics, as well as acoustic monitoring of sediment transport, will continue to utilize a variety of oceanographic instrumentation. Large research vessels will be needed to deploy and recover moorings and fleets of gliders. For these types of studies, future research vessels should be as acoustically quiet as possible (discussed in Chapter 4).

Research needs for physical oceanography in coastal environments are similar to those in the deep ocean, but are complicated by factors such as proximity to large river plumes, physiography of the continental shelf, and human activities that modify material fluxes from land and atmospheric deposition across broad shelf areas (Figure 2-1). The continental margin is more sensitive than the deep ocean to tidal and wind-driven circulation patterns, varying penetration of sunlight, and the interaction between river, estuary, and coastal zone runoff and upwelling events and eddy exchanges. Sample collection in coastal regions often occurs on smaller spatial and temporal scales than for deep ocean physical oceanography and is especially likely to require vessels with shallow drafts and excellent maneuverability and station keeping. Research needs for coastal currents and physical dynamics will be driven by advances in coastal ocean observing systems and associated sensors, and will be best met by more capable Regional/Coastal and Regional class vessels.

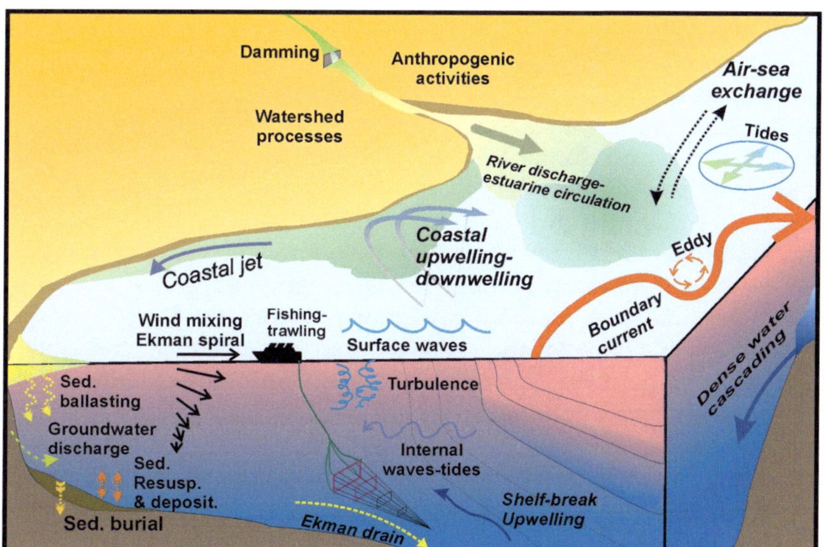

FIGURE 2-1 Important physical processes in continental margins (reprinted from Liu et al., 2009; with kind permission of Springer Science & Business Media).

CHEMICAL OCEANOGRAPHY

The field of chemical oceanography is directed toward understanding the distribution, transformations, and rates of cycling of the major and minor elements in the oceans. Major research issues driving current and future research include

- The ocean's role in the global carbon cycle, including oxygen and nutrient budgets that control biological productivity;
- Ocean cycling of climate-active gases (greenhouse gases, aerosol precursors, and stratospheric ozone-depleting substances);
- Ocean acidification resulting from ocean uptake of CO_2 and other anthropogenic emissions; and
- Quantifying fluxes of trace elements and isotopes into the ocean and developing an accurate understanding of the processes controlling their distributions.

Research needs include determining the regional and seasonal distributions of macro- and micronutrients that regulate primary productivity and influence ecosystem structure in the oceans, and characterizing the reservoir of dissolved and particulate organic carbon to understand its origin, cycling, and fate. Future research will encompass large-scale global ocean surveys (e.g., GEOTRACES; GEOTRACES Planning Group, 2006) and multidisciplinary regional process studies that interpret and constrain understanding of the paleoceanographic record through identification and quantification of chemical fluxes into the oceans and by developing greater understanding of the tracer potential of materials such as trace elements and isotopes. These types of research will require Global class ships that are capable of globally ranging, multi-investigator cruises with facilities that include adequate clean laboratory space and berthing accommodations for a large science party.

With the exception of salinity measurements, temporal and spatial chemical variability in the oceans is poorly documented, even for macronutrients such as nitrate and silicate. Almost all chemical and isotopic analyses cannot be done remotely and require ship-based hydrographic water sampling and shipboard or post-cruise laboratory work. Therefore, a primary driver for chemical oceanography cruises is clean laboratory space and an underway scientific seawater supply. A variety of new analytical approaches have greatly improved the capability to provide molecular analyses of carbonaceous material in the oceans (i.e., liquid chromatography-mass spectrometry, Fourier transform ion cyclotron resonance mass spectrometry). Exploitation of these and other advanced analytical techniques will spur the demand for ship-based water sampling for the foreseeable future.

BOX 2-2
Open Ocean Blooms in the North Pacific

Satellite ocean color observations have revealed large ephemeral open ocean plankton blooms (Wilson, 2003), but their origin, dominant species, and impact on biogeochemical cycles remain unresolved. In the following hypothetical future scenario, scientists investigate whether these blooms are intense sites of carbon export and how they affect food web dynamics. Although these blooms have been observed regularly in the North Pacific during the same season, scientists have not yet determined if they export particulate organic carbon (POC) to the deep sea or if it is instead remineralized near the sea surface. Routine glider patrols and Argo float data could indicate increased eddy activity, one characteristic of a potential incipient plankton bloom. After analyzing these data, scientists could schedule a cruise onboard an Ocean class vessel with a fleet of autonomous platforms, utilizing broadband communications to receive regular updates of satellite and modeling data regarding bloom development. Upon arrival at the site, a fleet of smart platforms would be deployed to track bloom expansion and movement. A ship-based, semi-automated command system will integrate in situ, remote sensing, and model information to intelligently navigate the AUVs. Scientists could then use detailed information from the AUVs to carefully target water sample collection, deploy drifting sediment traps, and launch net hauls to characterize the plankton bloom's impact on the food web and on element biogeochemical cycling. Researchers could use the information collected during the event to inform and refine global models.

The development of in situ chemical sensors for oceanography is an active area of research (Buffle and Horvai, 2000; Varney, 2000). Devices include electrochemical and colorimetric sensors capable of detecting various gases (oxygen [O_2], nitrogen [N_2], CO_2) and nutrients (nitrate, silicate), and mass spectrometers for trace gases (methane [CH_4]). These types of sensors are increasingly utilized on limited-duration drifters and buoys, and it is expected that in the future these sensors will be incorporated into large-scale observational systems such as the Argo float array. Such sensor networks are needed to detect changes in ocean chemistry associated with climate change. Before sensors can be used in this manner, stringent requirements for calibration and stability and long duration deployment will have to be addressed. These sensors will, however, provide only a small subset of the chemical, isotopic, and kinetic parameters that need to be measured to achieve a process-level understanding of the factors controlling seawater chemistry. Although new sensors are currently under development for the detection of a full range of chemical tracers, a majority of geochemical work in the foreseeable future is likely to be limited to shipboard sampling and analysis.

Satellite-based measurements of ocean color, sea surface temperature, and sea surface height have proven increasingly important to chemical, biological, and physical oceanography as a guide for process-oriented field studies (see Box 2-2) and as a basis for extrapolating in situ chemical measurements (McGillicuddy, 2009). A new space-borne ocean salinity sensor is scheduled for launch in late 2009 (European Space Agency, 2009). Satellite detection of specific chemicals (such as nutrients) in the oceans is not currently achievable and is unlikely to play a significant role in the near future.

Global carbon cycle research will also need to focus on continental margins. Future research will focus on biogeochemical processes along coastal margins, offshore particulate fluxes, sediment dynamics, and interactions between benthic and pelagic processes. While data collected from coastal observing systems will help to quantify carbon sources and sinks in this region, Regional/Coastal and Regional class vessels capable of working in the nearshore can collect a greater variety and volume of sediment, biological, and water samples in areas that are difficult to access using stationary or autonomous instruments, especially in response to unpredictable events.

ATMOSPHERIC CHEMISTRY AND AIR-SEA EXCHANGE

The exchange of trace gases and aerosols between the ocean and the atmosphere exerts a major influence on the composition, reactivity, and radiative properties of the atmosphere. Major research themes in atmospheric chemistry involving the oceans include the following:

- Tropospheric and stratospheric photochemistry
- Direct and indirect aerosol radiative forcing
- Atmospheric deposition of aerosol-borne nutrients derived from desert dust or anthropogenic pollutants.

These issues represent significant uncertainties in climate forcing and feedback and are poorly parameterized in the current generation of climate models. Research needs include determining the saturation state of many trace gases in the surface ocean, assessing the reaction and pathways of climate-active gases, and characterizing the composition, physical properties, and depositional patterns of aerosols (Lambin et al., 1999; Liss et al., 2004).

Atmospheric chemistry has some unique observational challenges because of the sensitivity of the atmosphere to a wide range of trace-level chemicals and because atmospheric transport and mixing are so rapid. Progress in marine atmospheric chemistry is observationally limited, with the need for broad spatial and temporal coverage as well as detailed in situ process studies. Future research will involve a combination of airborne and ocean-borne research platforms, with increasing use of unmanned aircraft and drones in conjunction with research vessels and buoys. The academic fleet will continue to play a unique and essential role in atmospheric chemistry research programs because it provides access to the marine atmosphere with a duration and payload unmatched by other platforms. Research vessels will also continue to play an important role as a test bed for new analytical instruments and as a platform for calibration and validation of the next generation of satellite-based atmospheric chemistry instruments.

The air-sea interface is a complex region that controls the transfer of heat, momentum, gases, and aerosols between the ocean and the atmosphere. Processes controlling air-sea exchange span a wide range of scales from the sub-millimeter thickness of the sea surface microlayer to the basin-wide scale of major ocean currents and atmospheric circulation systems. The interface is physically, biologically, and chemically complex. No adequate theoretical framework exists to describe transport across the air-sea interface, and the conditions occurring in nature cannot be replicated easily in the laboratory. Future research will focus on the development of physically based parameterizations of air-sea fluxes for use in regional and global climate models. This will require in situ observational process studies, involving general purpose and specialized research vessels as well as air-sea interaction buoys and aircraft. In situ process studies of air-sea gas and aerosol transfer have increased in both size and complexity, involving a variety of techniques including deliberate inert gas tracers, eddy covariance flux measurements, and passive and active sensing of the sea surface. Upscaling of fluxes from local to regional and global scales will involve buoys, satellite measurements, and modeling. The need for ship support of air-sea interaction studies is likely to increase in the future to carry out process studies, to support regional air-sea buoy networks, and to validate satellite-based measurements.

BIOLOGICAL OCEANOGRAPHY

Biological oceanography focuses on marine organisms and their relationship to ocean circulation, nutrients, and the biogeochemical cycling of elements. Emerging research issues in this field include

- Global biogeochemical cycles,
- Organisms' role in and response to climate change,
- Linkages between marine ecosystems and human health, and
- The dynamics and basin-wide connectivity of marine populations.

The biological pump plays an important role in the concentration of atmospheric CO_2. Current approaches to studies of the biological pump marry ship-based observations with autonomous systems, fixed observatories, and remote sensing and modeling. Newly developed genetic tools are presently used to examine the composition and function of marine microbes at the base of the food web, and they are being used to identify species of zooplankton and fish and their population structure (Bucklin, 2000; Scholin et al., 2009). In the future, these genetic tools (and others, including new biogeochemical sensors) will be adapted for use on autonomous platforms, ocean observatories, and systems such as Argo, and will lead to worldwide, data-rich measurements of the organisms that drive biogeochemical cycles. However, in the near future, research vessels will still be required to collect water and organisms for biological oceanographic studies. There will be a continued need for sustained, established time series.

Coastal ecosystems and human health are inextricably linked, yet these ecosystems are increasingly threatened by nutrient pollution, toxic bacteria and algae, and resource exploitation (e.g., Bank et al., 2007; Chen et al., 2008). Marine issues that directly impact human health and economics, such as harmful algal blooms (HABs), will require multidisciplinary, focused process experiments to better understand how these events occur (see Box 2-3). Since these issues tend to be near coastal margins, Regional

BOX 2-3
Nearshore Harmful Algal Blooms

In the near future, environmental observations from coastal moorings, regular glider patrols, Argo floats, and satellite data from the ocean surface will be fed into physical-biological models to monitor possible harmful algal bloom developments. With information from the models, a Regional class vessel can be used to deploy a fleet of AUVs and small moorings in a nearshore region where red tides are known to initiate, presumably by the resuspension of resting stages from the sediments into a growth environment. Once the resuspension event is detected, a shore-based command system that integrates observations, satellite data, and model information will instruct the AUVs to map the area with in situ sensors and to collect samples. Scientists aboard the ship can confirm that the samples contain target species, launch drifters to track the location of the HAB, and monitor the patch until it decays. Data collected during the experiment will provide a foundation for refining models and identifying critical observations to improve HAB prediction and mitigation.

or Ocean class vessels capable of engaging in multi-investigator, complex experiments are needed. Smaller vessels will also be needed to study biogeochemical processes in shallow coastal waters. In the future, these nearshore, ship-based programs will be complemented by an array of sensors mounted on moorings and observatories.

Ocean acidification is an important, growing area of research. Approximately 40 percent of the CO_2 introduced into the atmosphere from burning fossil fuels is being taken up by the ocean, affecting inorganic carbon equilibration and decreasing pH in the surface ocean (Sabine et al., 2004). It is currently unknown how this issue will affect ocean biodiversity, ecosystem structure, productivity, and the dynamics of marine populations. Ship-based research will be necessary to determine how much carbon is being taken up, where it is being stored, and how it is impacting marine ecosystems. Much of this research can be done on Regional/Coastal and Regional class ships, while ocean basin-scale studies will require larger Ocean or Global class ships, particularly in regions with higher sea states (e.g., Labrador Sea). In shallower waters, Regional/Coastal vessels will continue to be needed to study the impact of ocean acidification on coral reefs.

The role of climate change and anthropogenic pressures on marine populations and biodiversity is a crucial question. Programs such as Global Ocean Ecosystem Dynamics (GLOBEC) attempt to understand long-term, basin-scale variations in marine ecosystems through a combination of process studies and food web modeling. In addition to sophisticated modeling efforts and ocean observatories with continuous data collection, technological drivers include acoustically quiet instrumentation and vessels that are able to effectively conduct fish stock and mammal research. These types of programs often require the use of multiple ships in different regions during overlapping time periods.

Biological oceanography will continue to require manned submersibles and remotely operated vehicles for observation and sampling of deep sea and water column biota. These platforms are critical for sampling pelagic organisms in the midwater and for collecting organisms at the seafloor, including hydrothermal vent and methane seep communities and deep water corals. These communities are poorly known and currently undersampled. At present the deep submergence community is in the process of replacing the human-occupied submersible *Alvin* and has developed a new hybrid remotely operated vehicle (ROV) for exploration of deep subduction zone trenches (Bowen et al., 2008). In 2004, a National Research Council (NRC) report recognized the need for a second ROV (National Research Council, 2004). Since submersible and ROV crews take up a significant number of science berths and require large deck areas, there will be considerable pressure to conduct these cruises from Global class ships in the future.

MARINE GEOLOGY

Marine geology and geophysics (MGG) focuses on processes leading to the formation and evolution of the ocean crust and continental margins and their linkages to processes in the Earth's oceans, mantle, and continents. Although a significant component of research in MGG still involves exploration (National Research Council, 2003b), the past few decades have seen a general trend toward studies that integrate multidisciplinary observations to understand complex systems. Major research areas include the following:

- Paleoceanography and paleoclimatology
- Formation, evolution, and destruction of oceanic crust and lithosphere
- Sedimentary processes on continental margins
- Role of crustal fluids in the geologic cycle (i.e., crustal alteration, hydrothermal systems and chemosynthetic life, earthquakes)
- Geohazards, including tsunami generation and gas hydrates.

There is increased emphasis on MGG topics that are of immediate societal interest. Geological records are used in paleoceanography and paleoclimatology to understand processes that affect climate change and cause variability in the climate-ocean system on various time scales. Studies of gas hydrate deposits on the continental margin are motivated by their potential as an energy source and by their role in carbon sequestration. The 2004 Southeast Asian tsunami led to renewed studies of subduction zone earthquakes and the mechanisms of tsunami generation. Human occupation of the coastal zone drives continued need for process studies related to sediment resuspension and transport at the land-ocean interface and along the coastal margin.

While sophisticated laboratory measurements and computer modeling play an increasingly important role in MGG, research vessels remain an essential driver to explore the seafloor and underlying geological structures, sample rocks and fluids on and below the seafloor, and deploy seafloor instruments. Nearly every major research direction within MGG is at the forefront of utilizing new or improving observational and sampling technologies and thus requires research ships that are capable of using or deploying them.

Seafloor mapping, a prerequisite for many multidisciplinary oceanographic studies (see Box 2-4), serves both as an exploratory tool and as

BOX 2-4
Carbon Cycling Observatories in the Arctic

In this hypothetical example, a cabled observatory that has been deployed from Barrow, Alaska, to passively monitor bowhead whale migrations and methane seeps, as well as measure changes in the base of the ice canopy throughout the year using upward-looking sonars, is being expanded. The new observatory node will extend from the shelf to the base of the slope to measure organic carbon exchange in a changing Arctic Ocean. Placement of the new node will depend on identifying preferred sediment pathways by mapping regions where organic carbon is being deposited. Large-scale geomorphic features such as Barrow Canyon as well as smaller-scale features such as grooves and iceberg scours created by glaciogenic processes influence sediment pathways and deposition rates. A late-summer field geophysical mapping program will systematically collect multibeam and subbottom data for the region and then deploy an ROV for higher-resolution surveying and sampling of candidate sites. The ROV video system and sensors will simultaneously measure nutrients in the water and map macrofaunal benthic communities that could be affected along proposed cable pathways. Sediment samples will be collected and analyzed to estimate sedimentation rates for organic carbon using ^{210}Pb and ^{137}Cs. Satellite-derived images of ocean and ice canopy conditions will be beamed to the ice-strengthened vessel several times per day to provide advance warning of changing conditions and promote safe operations.

a means to answer basic science questions about seafloor processes. All Global and Ocean class ships in the UNOLS fleet are equipped with deep-water multibeam systems for regional-scale mapping. AUVs are effective platforms for collecting high-resolution bathymetric, sidescan, magnetic, and gravity data and have been demonstrated to successfully complement ship-based surveying for mapping at nested scales or as an add-on to cruises with other objectives.

Seismic techniques are a critical component of MGG studies along active plate boundaries and rifted continental margins. In 2008, the University-National Oceanographic Laboratory System (UNOLS) fleet upgraded its capacity to accomplish seismic surveying with the *Marcus Langseth,* which is capable of deep penetration two- and three-dimensional multichannel reflection profiling. The air gun seismic source can also be used for ocean bottom seismometer refraction experiments. Given the importance of seismic imaging to MGG and ongoing developments of new computational tools and interpretive capabilities, it is likely that MGG will place high demands on the *Marcus Langseth* and other platforms capable of carrying out seismic surveys.

Ocean bottom seismometers networks are increasingly being deployed for passive experiments to monitor microearthquakes along plate boundaries and to image lithospheric and upper mantle structure using teleseismic earthquakes. Plans for research in ocean mantle geodynamics (Oceanic Mantle Dynamics Workshop, 2002) envision an increased use of passive seismic arrays as well as the use of passive and active source electromagnetic experiments in combination with petrological and geochemical studies to understand the coupling between mantle convection and plate tectonics. All of these studies will require the largest ships in the UNOLS fleet for instrument deployment.

Deep drilling provides a means to sample sedimentary and igneous rocks and pore fluids below the seafloor, to measure physical properties below the seafloor, and to monitor hydrological processes. In sedimentary environments, coring techniques complement drilling and are important for coastal and paleoceanographic studies. At present, the Integrated Ocean Drilling Program (IODP) provides riser and riserless platforms to support drilling, while the 2007 commissioning of the Woods Hole Oceanographic Institution (WHOI) long corer on the *Knorr* (Figure 2-2) provides the capability to collect cores up to 45 meters long (Curry et al., 2008) and represents a critical, heavily utilized tool for the paleoceanographic community. Despite significant community interest (Sager et al., 2003), the U.S. academic fleet currently has very limited capabilities for drilling short holes robotically in igneous and lithified sedimentary rocks. The development of such a system is anticipated to increase the demand for platforms that can carry out drilling and coring.

MGG utilizes submersibles and ROVs for observational seafloor stud-

FIGURE 2-2 The WHOI long corer system mounted on the *R/V Knorr* (used with permission from James Broda, Woods Hole Oceanographic Institution).

BOX 2-5
Volcanic Eruption on the Juan de Fuca Ridge

In this hypothetical near-future scenario, the Ocean Observatories Initiative (OOI) regional cabled observatory deployed on the Juan de Fuca plate and the U.S. Navy's Sound Surveillance System (SOSUS) hydrophone network detect a three-day swarm of earthquakes with the signature of a volcanic eruption on the southern Juan de Fuca Ridge. At the same time, sensors deployed around the nearest cabled node, 150 km to the north on Axial Seamount, detect high rates of local microearthquakes and increased fluid temperatures and flow rates in nearby hydrothermal vents. The mid-ocean ridge community mounts an event response cruise to better understand impacts of volcanic eruptions on chemosynthetic biological communities, resolve a long-standing controversy on the origin of event plumes (hydrothermal plumes that rise 1 km above the seafloor soon after an eruption), and understand how a volcanic eruption triggers changes in seismicity and hydrothermal flow 150 km away. Within a week, a Global class ship equipped with an ROV reaches the eruption site. Telepresence allows several key scientists to participate from shore and shipboard scientists to stream live video of ROV operations to science museums and aquariums across North America. Despite challenging weather conditions, the scientists are able to explore the eruption site, collecting rocks and hydrothermal fluid samples for chemical and microbial analysis. Between ROV dives, the shipboard CTD detects and samples an event plume. Scientists are able to launch floats into the event plume, tracking its movements for future sampling. The ship then transits to the Axial Seamount cabled node, where the ROV replaces fluid samplers and sensors that have failed. Using samples collected during the cruise, the science party initiates a discussion with funding agencies regarding the feasibility of a follow-up cruise in 2 to 3 months.

ies. They are necessary for multidisciplinary studies of hydrothermal systems (see Box 2-5), creating detailed geological maps, precise rock sampling and coring in complex terrain. In addition, they will continue to be essential for servicing instruments—for example, to monitor fluid pressure, temperature, and chemistry in Ocean Drilling Program boreholes.

The Ocean Observatories Initiative (OOI) has been motivated by the recognition that sustained time series observations are critical to many fields of oceanography. In MGG, observatories are necessary to characterize volcanic eruptions and large earthquakes and to monitor their impacts on fluid discharge across the seafloor and chemosynthetic biological communities. These observatories will require significant Ocean and/or Global class ship time, with an ROV that is capable of deploying short runs of thin cable, junction boxes, and a wide variety of sensors.

OCEANOGRAPHY EDUCATION AND TRAINING

Addressing the future ocean sciences research agenda will require a cadre of well-trained seagoing scientists. Students need to gain experience and training at sea to become scientists that are well versed in the broad field of oceanography. Gaining experience at sea is nearly as crucial for future oceanographers who will do their work ashore as it is for those who will run ship-based research experiments, in whose case at-sea experience amounts to a type of apprenticeship. New technologies will enhance education on shore but are unlikely to change the paradigm. The academic fleet will need ships with sufficient berthing to carry not only the science and technical teams, but also the next generation of oceanographers.

CONCLUSIONS

The future ocean sciences research agenda will be driven by a diverse portfolio of disciplinary and interdisciplinary seagoing studies across a broad range of spatial and temporal observational scales. The fleet of the future will be required to support increasingly complex, multidisciplinary, multi-investigator research projects using autonomous technologies, ocean observing systems, remote sensing, and modeling. Research vessels will be needed to investigate and explore all areas of the ocean, from tidal zones to deep trenches.

Recent advances in technology (such as global arrays of floats and satellite data) have fundamentally altered oceanographic research, with sampling coverage and frequency that far outweigh the collection abilities of the research fleet. However, several new technologies that will impact future ocean research (e.g., in situ chemical and genetic sensors) have not yet been proven capable of withstanding the rigors of deployment on a mooring, autonomous vehicle, or ocean observing system, and most of these systems will require both ship deployment and calibration. In the next 10-20 years, autonomous mobile platforms and fixed observatories are not expected to have sufficient sensing capabilities to replace traditional research vessels.

Ship-based measurements will continue to be required in the foreseeable future to further both basic research and new discoveries in the ocean. A capable academic research fleet will continue to be required for needs such as water sampling, calibration and validation of satellite remote sensors, seafloor mapping and drilling, focused process studies, and atmospheric sampling. As continuous ocean observing systems and future generations of autonomous and fixed platforms document novel phenomena and processes in the ocean environment, they are likely to drive increased demand for ship time to research these new discoveries further. Ships will also continue to be required to train students and advance the study of oceanography.

3

Technological Advances and Their Impact on the Fleet

How technological advances such as autonomous underwater vehicles and ocean observing systems will affect the role and characteristics of the future UNOLS fleet with regard to accomplishing national oceanographic data collection objectives.

How evolving modeling and remote sensing technologies will impact the balance between various research operations such as ground-truthing, hypothesis testing, exploration, and observation.

The pace of technological advances in oceanography continues to accelerate, and these changes fundamentally alter how science is accomplished. Enabling technologies considered essential in 2009—such as the Global Positioning System (GPS), satellite communications and the Internet, remotely operated vehicles (ROVs), autonomous platforms, and sensors such as multibeam mapping systems and improved vessel-deployed chemical sensors—were in their nascent stages or did not exist when many vessels in the academic fleet were built. Proven cutting-edge technologies are often adopted rapidly by the scientific community, resulting in post-fabrication modification of the research fleet that can be costly and provide less than optimal performance. In this chapter, recent technological advances for ships and shipboard support of science are reviewed and their likely impact on the future oceanographic research fleet is discussed. Technologies considered include dynamic positioning systems, aloft sensors, satellite systems, long coring systems, autonomous underwater and airborne vehicles, remotely operated vehicles, ship-to-shore communications and telepresence, and ocean observing systems. As the number and complexity of seagoing systems increase, so does the need for broadly trained and highly skilled technicians to maintain them, a topic covered briefly at the conclusion of this chapter.

DYNAMIC POSITIONING

One of the technologies already in use on many oceanographic research ships is dynamic positioning (DP). By properly controlling bow thrusters, azimuthing propulsers, and other elements of a ship's propulsion system, DP makes it possible for a ship to hold a given geographical location and required heading even under severe conditions. DP also contributes to operation quality and efficiency because waypoints can be used to minimize time between stations and heading and ship track can be controlled accurately over long distances. The maneuvering and propulsion system is linked to the ship's navigation system to ensure the position is fixed. DP systems installed retroactively on vessels sometimes have inadequate propulsion and computer systems to maintain station in high sea states. Newer systems utilize onboard computers to control the machinery. Nine University-National Oceanographic Laboratory System (UNOLS) vessels currently have DP systems (all Global and Ocean class, the Intermediate *Seward Johnson*, and the Regional/Coastal *Hugh R. Sharp*). Of these, *Knorr* and *Melville* had systems installed retroactively, and *Revelle*, *Atlantis*, *Thompson*, and *Marcus Langseth* have had replacement systems installed (Annette DeSilva, personal communication, 2009). With the growth of offboard vehicles and the need to safely deploy and recover these systems, it is expected that DP will become a standard feature of research vessels rather than a special case. The Navy's systems specifications for the planned Ocean class vessels explicitly states that a DP must be installed (Naval Sea Systems Command, 2009).

ALOFT SYSTEMS

Aloft systems include instruments such as meteorological sensors, GPS and communications antennas, and instruments for measuring ocean surface reflectance. At present the upper portions of research vessels are not designed so that all aloft systems have the appropriate exposure; rather, they compete for space in a crowded part of the vessel and performance is compromised. The vessel requirements for GPS and communications systems, for example, concern clear sight lines between the antennas and the required satellites in any possible position, from overhead to the horizon in any direction. In current installations, the antennas can be blocked on certain headings by the vessel stack. Likewise, solar references for satellite calibration need to be mounted so that interference from ship shadows and light reflecting off the hull are minimized (Hooker, 2009). Future vessels designed with these specifications in mind should have sufficient space aloft to accommodate all atmospheric and oceanographic sensors as well as navigation and communications satellites without mutual interference.

SATELLITE SYSTEMS

The oceanographic community currently utilizes satellite-based measurements of ocean color, sea surface height, sea surface temperature, and surface winds to characterize ocean variability and to study physical and biogeochemical processes. The use of remotely sensed data is expected to grow as new satellite-based instruments are deployed and new generations of ocean models improve our ability to integrate satellite and in situ observational data. The future research fleet will require increased bandwidth to relay large satellite datasets between ship and shore (discussed in greater detail later in this chapter) and additional capabilities for ship-based calibration of space-borne instruments (e.g., the solar references mentioned above).

LONG CORING

The collection of sediment piston cores with lengths ≥40 meters is an increasingly important technique for paleoceanographic and continental margin studies. Long coring capability was added to the UNOLS fleet in 2007 through installation on the Global class, 279 foot (85 meter) *Knorr* (see Figure 2-2), which is now able to collect cores up to 46 meters in length (Curry et al., 2008). To accommodate the weight of the coring system, the deck was strengthened significantly, and a more robust A-frame and winch were added. Given that *Knorr* is slated to be replaced in 2015, the future fleet will need to plan for at least one vessel capable of reliable, safe collection of long cores. Increasing scientific demands for long coring operations throughout the world could lead to demand for coring systems on more than one academic research ship.

AUTONOMOUS VEHICLES

Autonomous systems are becoming increasingly available and have found many applications to a variety of scientific problems. In many cases these systems have transitioned to operational status and have received wide acceptance by the scientific community. Examples include floats

(Davis, 1991; D'Asaro, 2003), gliders (Stommel, 1989; Davis et al., 2003; Rudnick et al., 2004), autonomous underwater vehicles (AUVs; Yoerger et al., 1998), and unmanned aerial vehicles (UAVs). The application of these systems should expand greatly in coming years as their capabilities improve and their application to science problems becomes better understood.

Enhanced capabilities will cover a spectrum from longer-range capabilities as energy sources and vehicle efficiencies improve to smaller-scale applications as the utility of microsystems such as micro-AUVs is demonstrated. New and varied in situ sensors including mass spectrometers and genomic sensors, sampling systems designed for autonomous operation, and improvements in overall reliability will also increase demand for autonomous systems. The range of commercially available vehicles will certainly expand, allowing vehicles to be better matched to specific science problems. Operational groups will likely become more comfortable with increasingly aggressive deployment strategies, particularly the use of multiple vehicles in the water at the same time with unattended operation. Autonomous platforms and their associated launch and recovery systems will come in many different forms, many of which will place new and varied demands on oceanographic vessels.

It is worth noting that continuing advances in battery life and adaptive programming will lead to a greater potential for launching autonomous vehicles (e.g., gliders) directly from the nearshore, whether in small boats or from the beach itself. In this way, missions involving autonomous platforms might be independent of research vessels for launch and recovery. However, this capability is still in its developmental stages, and for many locations, rough topography and/or wave and current regimes will discourage launching directly from the beach in the foreseeable future.

Floats and Gliders

Autonomous floats are a mature technology specifically designed for easy deployment in great numbers. For instance, the Argo program (*http://www.argo.ucsd.edu/*), part of the Global Ocean Observing System, has more than 3000 floats profiling throughout the ocean on a continuous basis (Figure 3-1) and provides an unprecedented view of the upper ocean's circulation and hydrography. Most floats are expendable so they make no special demands on vessels and can be deployed from ships of opportunity. Future advances will increase the number and type of sensors that are carried on floats (i.e., chemical, biological), strengthening demand for ship-based water sampling for sensor calibration. Gliders are significantly more expensive and are often recovered, although Navy-funded developments are in progress for surface ship- or air-deployed

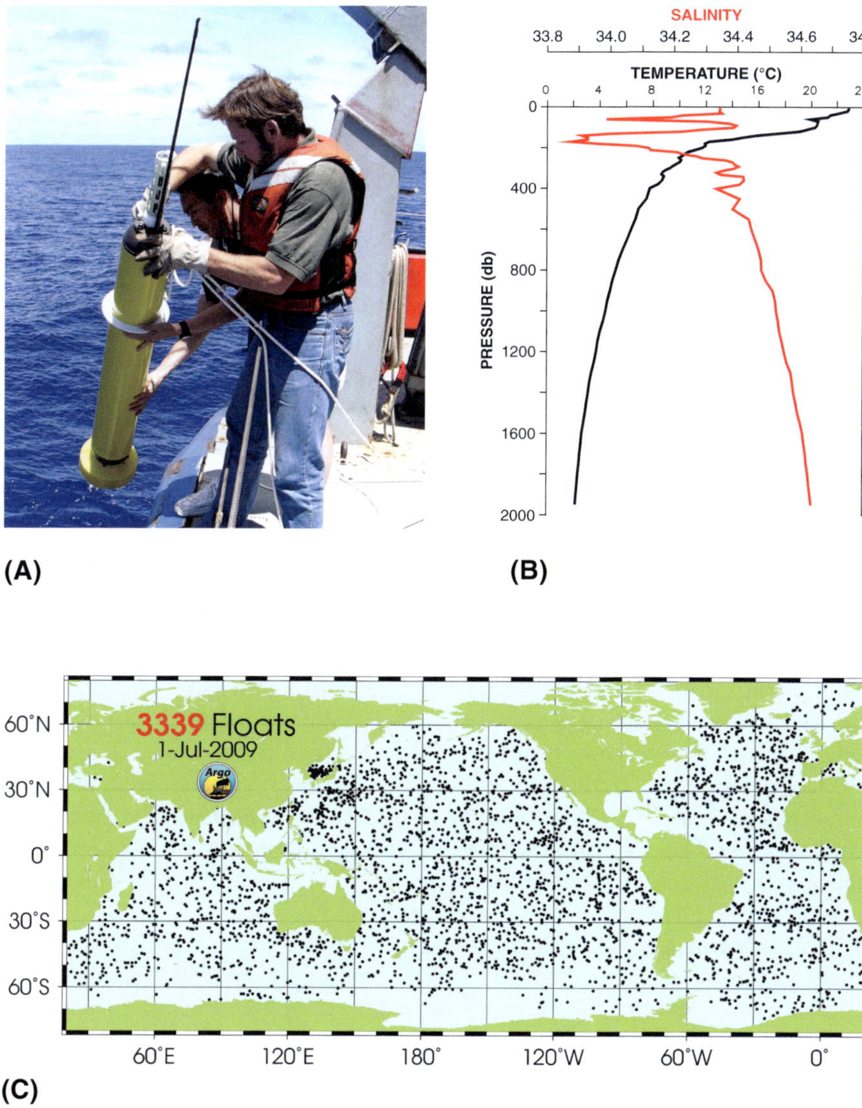

FIGURE 3-1 (A) An Argo float being deployed in the North Pacific Ocean with the *Melville* in 2004 (used with permission from James Swift, Scripps Institution of Oceanography). (B) An Argo profile from the subtropical North Pacific. Temperature (black) and salinity (red) are shown. (C) The Argo float array in July 2009. Each black dot represents a float that has returned data within the last 30 days (B and C used with permission from Argo Project Office, *http://www.argo.ucsd.edu*).

expendable gliders (i.e., 2008 Navy Small Business Technology Transfer solicitation #N08-T016). The expendable gliders will likely have minimal sensor capabilities, however, and the continued use of fully configured, recoverable gliders is anticipated. Because of their low drag shape and minimal buoyancy when surfaced, gliders are difficult to recover. Their recovery is quite sensitive to weather conditions because of their low visibility on the surface and their potential for collision with the ship when they are hauled aboard. Ship design trends that facilitate the use of gliders includes lower freeboard, better over-the-side (OTS) handling systems, and acoustic and/or optical technology to assist with spotting vehicles on the surface. These changes will also benefit AUVs, discussed in the next section.

Autonomous Underwater Vehicles

Typical tasks for present-day AUVs include high-resolution seafloor mapping and measuring oceanographic phenomena such as temperature and salinity anomalies on spatial scales on the order of hundreds of kilometers over time scales of several days to perhaps weeks. With the advent of submerged docking stations (described in the section on ocean observatories), AUV duration limits will effectively be removed for areas with the required infrastructure. However, because docking stations will require fixed infrastructure, continued use of survey AUVs in an expeditionary mode (where they are launched and recovered for each battery charge) is expected. Advances in AUV technology are pushing toward both ends of the size scale, with very large AUVs (Tangirala and Dzielski, 2007) proposed to conduct basin-wide surveys over longer periods and micro AUVs[1] potentially hibernating at sites of suspected pending activity to facilitate extremely rapid event response.

The level of autonomy for AUV operations should increase significantly in the near future. Survey AUVs are usually operated today with continuous monitoring from a surface vessel. In some cases, the presence of the surface vessel is required for updating the vehicle's navigation system; in other cases the vessel monitors sensor data quality and remains in the vicinity should the vehicle surface early due to an unexpected fault. In many cases, the high cost of the vehicle combined with the possibility of problems makes continuous monitoring prudent. This situation is certain to change as navigation techniques evolve and operational confidence improves. When vehicle operations have reached a level of maturity that does not include continuous monitoring, oceanographic vessels will be needed to service fleets of AUVs. The size of the AUV fleets will be limited

[1] *http://robotik.dfki-bremen.de/de/forschung/projekte/unterwasserrobotik/uauv.html.*

by the ability of the vessel to deploy and recover vehicles continuously, quickly, and safely over a wide weather window.

Autonomous system operations will require ships that are equipped with specialized acoustic systems, lab space and berthing for operators, and launch and recovery of OTS handling gear. Acoustic systems used to track multiple vehicles using ultrashort baseline (USBL) navigation with integrated acoustic communications capabilities will be required for sophisticated multivehicle operations. These systems will likely become part of the vessel infrastructure and should not be adversely affected by noise radiated by the vessel. Safe and efficient launch and recovery of a variety of AUVs will also place demands on future vessel design. Specialized handling systems are and can be used with existing systems—for example, the deep water REMUS AUV—but one OTS handling system is unlikely to be compatible with all AUVs. The operation of multiple AUVs from a single vessel will require careful layout of deck space and may even require a different trade-off between deck and laboratory space. Furthermore, the deck used for AUV recovery, whether aft or amidships, would benefit from being closer to the waterline than it is on most current research vessels. AUVs are also likely to alter the composition of seagoing scientific teams with possible impact on lab space and berthing. Fleets of AUVs could generate very large datasets requiring teams of skilled personnel for processing; alternatively, the data processing requirement could be decreased by the ability to connect to shore via broadband communications.

Unmanned Aerial Vehicles

A relatively new technology for oceanographic research is the unmanned aerial vehicle. Most current UAVs are derived from recent military applications and are fairly expensive and complex (Winokur, 2009). As the technology becomes proven and adapted to the ocean environment, less expensive UAVs are likely to be used for research in remote areas and those with large areal extents. In 2009, the National Oceanic and Atmopsheric Administration (NOAA) used a UAV to monitor the location and distribution of seals in the Bering Sea.[2] In this case, the UAV was launched from a research vessel with a portable catapult and collected images and video before recovery with a catchline attached to a crane on deck. There are only brief mention of, and no current specifications for, UAVs in the UNOLS Science Mission Requirements for the Ocean and Regional class vessels (UNOLS Fleet Improvement Committee, 2003a,

[2] *http://alaskafisheries.noaa.gov/newsreleases/2009/aircraft060209.htm.*

2003b). As the use of ship-launched UAVs increases, launch and recovery options are likely to be factored into future ship designs.

REMOTELY OPERATED VEHICLES

Remotely operated vehicles have been used to conduct oceanographic research since the 1960s. They are used for a variety of purposes, including water, rock, and biological sampling; deployment and recovery of equipment; collection of still and video imagery; and seafloor mapping. ROVs have a number of requirements in common with their AUV counterparts, including OTS handling systems that allow safe and efficient launch and recovery as well as limited freeboard of the deck from which they are launched. In addition, because ROVs are attached to a ship via cable, they frequently require a specialized winch and wire system that accurately monitors the length of cable between the instrument and the vessel and can recover wire very quickly in the event unexpected entanglements are encountered. ROVs generally also need good ship DP in order to reliably navigate through treacherous terrain to acquire samples. Support teams for ROVs can be as large as AUV teams, so similar concerns about available lab space and berths apply. At present many research vessels can accommodate ROV operations without extensive modification, but use of these systems in the future would be improved by designing vessels that are more stable, with greater deck and lab space and more capable OTS launch and recovery systems. Future trends in ROV tools may follow the hybrid vehicle *Nereus*, which can operate as either an AUV for seafloor surveys or an ROV to collect samples (Bowen et al., 2008). An equally important trend will be robust ROVs that are capable of deploying and servicing heavy pieces of equipment and recovering large rock samples from the seafloor.

SHIP-TO-SHORE COMMUNICATIONS AND TELEPRESENCE

Real-time satellite Internet connections currently play an increasingly important role in operation of the UNOLS fleet and are expected to become even more significant in the future. At present, the larger ships in the fleet are equipped with the HiSeasNet system (*http://hiseasnet.ucsd.edu*), which provides shared connections at rates ranging from 64 to 256 kbps (kilobits per second) each way. Several UNOLS vessels are in the process of installing a system that will provide up to 432 kbps of additional bandwidth.

The availability of Internet connections on the UNOLS fleet serves several purposes. It contributes to science operations by allowing the exchange of data, models, and ideas between seagoing scientists and technicians and their colleagues ashore. Satellite observations and shore-based modeling of data collected aboard ship can be used to guide an experiment, and it is expected that this will occur with increasing sophistication and seamlessness in the near future. If complex instrumentation breaks down, satellite Internet connections allow shipboard technicians to interact with experts ashore to troubleshoot and make repairs. Internet availability also enhances educational and outreach activities by connecting the world to the ship through telecasts, web pages, and blogs. It provides scientists and crew with access to the web and personal email, improving the quality of life aboard the ship and playing a significant role in crew retention.

In 2005, several research cruises aboard UNOLS and NOAA vessels experimented with very high bandwidth connections that supported real-time digital video transmissions directly from an ROV to shore (i.e., Visions '05 [*http://www.visions05.washington.edu/*] and Lost City 2005 [*http://oceanexplorer.noaa.gov/explorations/05lostcity/welcome.html*]). In some cases, shore-based scientists sitting in a control room could participate in or even direct the exploration and sampling of the seafloor, while streaming live video to aquariums, museums, and schools served as a powerful education and public outreach tool. The NOAA ship *Okeanos Explorer* will make extensive use of such telepresence to engage shore-based scientists and the public in ocean exploration.

Within the UNOLS fleet the trend toward increasing bandwidth and decreasing costs of digital connectivity will likely influence science operations. However, it is unlikely to lead to decreasing demands for science berths. A typical science party includes personnel to control the experiment, run equipment, log operations, and process samples and data and provides berths to students who are receiving at-sea training and experience that is critical to their career development. As experiments become increasingly multidisciplinary and technically complex, the demands for science berths will increase. Similarly, scheduling that optimizes the use of ship time by supporting several experiments on a single leg will also increase the demand for science berths.

Viewed in this context, the emerging availability of a telepresence at sea provides a means to alleviate the pressure for science berths while enhancing the efficiency of operations. Although it is technically feasible to participate in science operations from a shore-based control center, it is difficult over the long term to balance the regular routine of shore-based life with the unpredictable 24-hour schedule of operations and decision making at sea. Instead, telepresence is likely to become a useful tool for involving shore-based scientists and technicians in intense components of a cruise that last only for a short duration, data analysis tasks that can be performed on a regular schedule, and troubleshooting of scientific equipment.

OCEAN OBSERVING SYSTEMS

The Ocean Observatories Initiative (OOI) is a National Science Foundation (NSF) contribution to national and international efforts for development of new long-term observing capabilities for the oceans. The OOI Science Plan (Daly et al., 2006) emerged from extensive community discussions (National Research Council, 2000b; Jahnke et al., 2002, 2003; Glenn and Dickey, 2003; Purdy et al., 2003; Schofield and Tivey, 2004; Daly et al., 2006) that were motivated by the recognition that many important processes occur over time scales and spatial domains that cannot be observed effectively using conventional ship-based expeditions or satellite observing platforms. The OOI aims to establish an interactive, globally distributed network of sensors in the oceans that will use pioneering technology to facilitate new research approaches. The system will have three components: (1) a global ocean observatory of highly capable moored buoys sited around the world's oceans, (2) a regional cabled ocean observatory that will instrument the seafloor and overlying ocean on the scale of a tectonic plate, and (3) a coastal observatory that will include both fixed and relocatable shallow water mooring arrays. These three field components will be integrated by a system-wide cyberinfrastructure that will allow scientists to access data in near real time and adapt their experiments to changing conditions.

The ship and deep sea submergence needs of the OOI were addressed in 2003 as part of a National Research Council (NRC) report on the implementation of ocean observatories (National Research Council, 2003a) and in a report prepared by a UNOLS Working Group (Chave et al., 2003). Since 2003, the design of the OOI has evolved considerably in the face of technical challenges and budgetary constraints. As a result, the ship time requirements are substantially less than initially envisioned. In the current plan (data from NSF, 2009), the global component is composed of arrays of three to four moorings and accompanying gliders deployed at four sites: the Southern Ocean southwest of Chile, the Irminger Sea southeast of Greenland, Station Papa in the Northeast Pacific, and the Argentine Basin. Approximately one month of Global class ship time per site will be required annually to install and service the global stations. The regional cabled component includes three science nodes on the Juan de Fuca plate and will require approximately two months of a Global class ship and ROV to service each year. The coastal component comprises a variety of moorings, gliders, and AUVs that will be deployed in the permanent Endurance Array off the coast of the northeast Pacific and in the moveable

Pioneer array first deployed in the Mid-Atlantic Bight. The coastal arrays will require approximately four months of combined Intermediate and Local ship time per year. In addition to these requirements, some specialized tasks may require the use of chartered vessels, and it is likely that other vessels (such as the Ocean or Regional/Coastal classes) will be used as needed, especially when the Intermediate vessels retire.

Although the NRC and UNOLS Working Group reports (Chave et al., 2003; National Research Council, 2003a) overestimated the ship time requirements of the OOI compared to its present scoping plan, many of their findings regarding the required capabilities of the ships still hold. UNOLS Global class ships were configured for programs such as the World Ocean Circulation Experiment (WOCE) and the Joint Global Ocean Flux Study (JGOFS) that emphasized fuel economy and cruise duration, large shipboard science parties, extensive laboratory space at the expense of deck space and limited heavy lifting in OTS operations. The needs of the OOI are significantly different. Installation and maintenance of OOI components would benefit from large deck spaces, the ability to lift and deploy heavy loads over the side, DP systems that can hold station in high latitudes and rough weather, the ability to have ROV operations, and the ability to store and install short lengths of cable.

Both the NRC and UNOLS Working Group reports (Chave et al., 2003; National Research Council, 2003a) noted that the current UNOLS fleet renewal plans do not adequately address the ship requirements of the OOI. In particular, they note that the new Ocean class vessels are not particularly well suited for ocean observatory operations. As discussed further in Chapter 4, the Science Mission Requirements (SMR) for Ocean class ships call for the ability to hold station in sea states up to 5, wind speeds up to 35 knots, and currents up to 2 knots. These specifications may not be sufficient for observatory purposes. In addition the SMR provides for only 1500-2000 square feet of aft deck space and winches and cranes that are similar to the current Global vessels and thus not well suited to heavy lifting. In addition the SMR calls for only 20-25 science berths, which may be inadequate for the long cruises to service buoys in remote locations or for housing the ROV, engineering, and science teams necessary for operations on the regional cabled observatory. However, response cruises or short repair cruises with an ROV could conceivably be staged with an Ocean class ship.

SEAGOING MARINE SCIENCE TECHNICIANS AND THEIR EVOLUTION

The preceding description of the rapidly evolving and highly technical systems for future oceanographic research vessels likewise will place evolving and technical demands on the personnel sent to sea to operate and maintain these systems. In the past, science technicians provided by UNOLS ships to support seagoing equipment focused on deck operations and operated a relatively small inventory of the ship's installed scientific equipment, such as echosounders. Science teams often brought their own experienced technicians to maintain and operate the equipment they brought aboard. Today, fewer seagoing scientists employ full-time technicians, and the array and complexity of both installed shipboard scientific equipment and user-supplied equipment has greatly expanded. As a result, shipboard science technicians must now play a variety of roles. They are liaisons between scientists and the ship's crew, educators, and communicators; support many different science and data systems (including user-supplied systems they may never have seen before); collect data in the absence of a principal investigator; and assist in cruise planning and logistics (Fisichella, 2009). UNOLS institutions are finding it hard to recruit qualified technical support with such broad experience.

In addition, the funded complement of shipboard technicians on UNOLS vessels is currently limited by supporting federal agencies, which has helped to slow growth in technical support costs. Ship operators, however, see the need for more, better-trained shipboard science technicians. The two shipboard technicians now carried on general purpose Global class vessels are a minimum for most cruises, and on many cruises they simply cannot attend fully to all of their assigned tasks.

Future trends regarding shipboard support indicate that both the increasing complexity of tasks and the shortfall of technical expertise will continue in the near future. Future tasks will include facilitating ship-to-shore communications; supporting more extensive AUV, UAV, and ROV operations; servicing ocean observatory sensors and infrastructure, managing and interpreting larger and more complex datasets; and supporting shore-based as well as shipboard needs. Limited berthing space and telepresence may also lead to technical personnel being tasked with data collection and/or instrument deployment in lieu of shipboard scientists. In this mode, technical expertise and training become critical to mission success. In addition, seagoing technicians will be responsible for the safe operation of simultaneous tasks and balancing constraints such as space and power requirements. If more technicians are needed for ship or equipment support in the future, there will be further demand to find highly qualified personnel. Sharing technical personnel between operating institutions may alleviate some of these issues, providing expertise and steady employment. However, this issue is unlikely to impact the design of future ships, with the exception of science berthing.

CONCLUSIONS

Technological advances in oceanographic sensors and platforms have enhanced the use of research vessels, allowing for vastly extended data collection at greater distances from the ship. Ocean observatories and autonomous vehicles will impact future vessel design requirements for acoustic communications, deck space, payload, berthing, launch and recovery, and stability but will not lessen the need for vessels themselves. Aloft sensors, especially those used for calibration of satellite data, will require high spaces with adequate lines of sight. There is need for increased ship-to-shore bandwidth, in order to facilitate real-time, shore-based modeling and data analysis in support of underway programs, allow more participation of shore-based scientists via telepresence, and increase opportunities for outreach. Dynamic positioning systems are very likely to become standard components of oceanographic research vessels to support increasing use of offboard vehicles that require precise positioning. Future research vessels will require improved over-the-side handling systems to facilitate deployment and recovery of instruments in higher sea states. Laboratory and deck spaces will increase in size, in order to allow deployment, recovery, and maintenance of large and technically complex instruments such as AUVs, ROVs, and large systems (e.g., moorings) that will support long-term ocean observing. Servicing ocean observatories and launching and recovering autonomous vehicles will result in increased demands for ship time. To support these systems and data, more highly qualified and trained seagoing technicians will be needed.

4

Oceanographic Research Vessel Design

The most important factors in oceanographic research vessel design. Does specialized research needs dominate the design criteria and, if so, what are the impacts on costs and overall availability?

Ship design is an exercise in conflict resolution. It is the creation of a system of systems to perform a specific mission while balancing conflicting requirements to achieve a ship capable of performing its mission in the best way possible within economic constraints. Oceanographic ship design is one of the very complex subsets of ship design, due to the large variety of oceanographic missions: physical, biological, and chemical oceanography; marine geology and geophysics; ocean engineering; and atmospheric science. Each discipline has its own unique set of mission requirements, yet a given ship is often called upon to perform work for a number of different disciplines, often on the same research cruise. In addition, the capital needed to build effective oceanographic ships is finite and scarce.

Ships will remain the primary method of conducting oceanographic research, both through direct observation and through deployment and recovery of sensors, moorings, and vehicles. Driven in part by national oceanographic research objectives, research will be conducted in increasingly remote and environmentally challenging areas. Future ships must be able to perform their science missions in all areas of the oceans, including the margins of the polar seas. Specialized vessels (icebreakers) will also be needed to work in ice-covered regions.

SCIENCE-DRIVEN SHIP DESIGN REQUIREMENTS

The future science trends and technology advances that will drive oceanographic ship design have been described in Chapters 2 and 3. These have been synthesized into a matrix (Table 4-1). Several of these needs are unique to certain disciplines and are potential design requirements that should be assessed carefully in general purpose oceanographic ship design. Other needs are more universal; for example, the ability to collect seawater samples throughout the water column is important for most of the oceanographic disciplines. Specific design considerations driven by the listed needs are discussed in the following sections.

Handling Equipment

Handling equipment overboard and onboard will continue to be of paramount importance, to allow for the safety of personnel, equipment, and the ship itself (Figure 4-1). Trends indicate that handling equipment must be able to operate effectively and safely up to sea state 6. General purpose oceanographic research ships require a permanently installed suite of winches (direct pull and traction) to perform conductivity-temperature-depth (CTD) type activities, deep tow, coring, and trawling missions. To expand the environmental operating window, active heave compensation has been incorporated on a number of recent ship designs. The Office of Naval Research (ONR) and the National Science Foundation (NSF) jointly funded a 2004 workshop to consider future handling systems.[1] Recommendations from that workshop were used in motion compensation systems installed on the Regional/Coastal class *Sharp* (Figure 4-1B,C), the Ocean class *Kilo Moana*, and the system designed for the Alaska Region Research Vessel (ARRV). It is likely that active heave compensation will be considered for all future University-National Oceanographic Laboratory System (UNOLS) vessels.

Gliders, autonomous underwater and unmanned aerial vehicles (AUVs and UAVs), and remotely operated vehicles (ROVs) often require specific deployment and recovery procedures and equipment (e.g., Figure 4-1A). Although systems vary, deployment is usually much easier than recovery. While UAVs now use catchlines for recovery, advancements in remote aircraft are likely to change significantly in the future. Current oceanographic vessels, especially the larger classes, have high freeboard that makes recovery more difficult for offboard equipment. Requirements for damage stability[2] and personnel safety in desired higher sea state

[1] *http://www.unols.org/publications/reports/lhsworkshop/index.html*

[2] Damage stability refers to the ability of a ship to have sufficient stability to survive a flooding casualty.

Table 4-1 Science-Driven Ship Needs

Science Driver	Physical	Biological	Chemical	MG&G	Atmospheric
Atmospheric measurement capability	X	X	X		X
AUV/glider/UAV stowage and handling	X	X	X	X	X
Capability to service observatories	X	X	X	X	
Clean laboratory space		X	X	X	X
Controlled temperature laboratory space		X	X		
Dynamic positioning	X	X	X	X	
High data rate communication	X	X	X	X	X
Hull mounted and deployable sensors[a]	X	X	X	X	X
Low radiated noise	X	X		X	
Low sonar self noise	X	X		X	
Manned submersible use		X	X	X	
Mooring/buoy deployment and recovery	X	X	X	X	X
Multi-channel seismics	X			X	
Ocean drilling and coring				X	
Precise navigation	X	X	X	X	X
ROV stowage and handling	X	X	X	X	
Towing nets and/or vehicles	X	X	X	X	X
Underway scientific seawater supply	X	X	X	X	X
Watercatching/water column sampling	X	X	X	X[b]	X

[a]In this instance, deployable sensors include centerboards, stalks, and towed sensors that can be lowered beneath the level of bubble sweep-down interference.

[b]For hydrothermal plume studies.

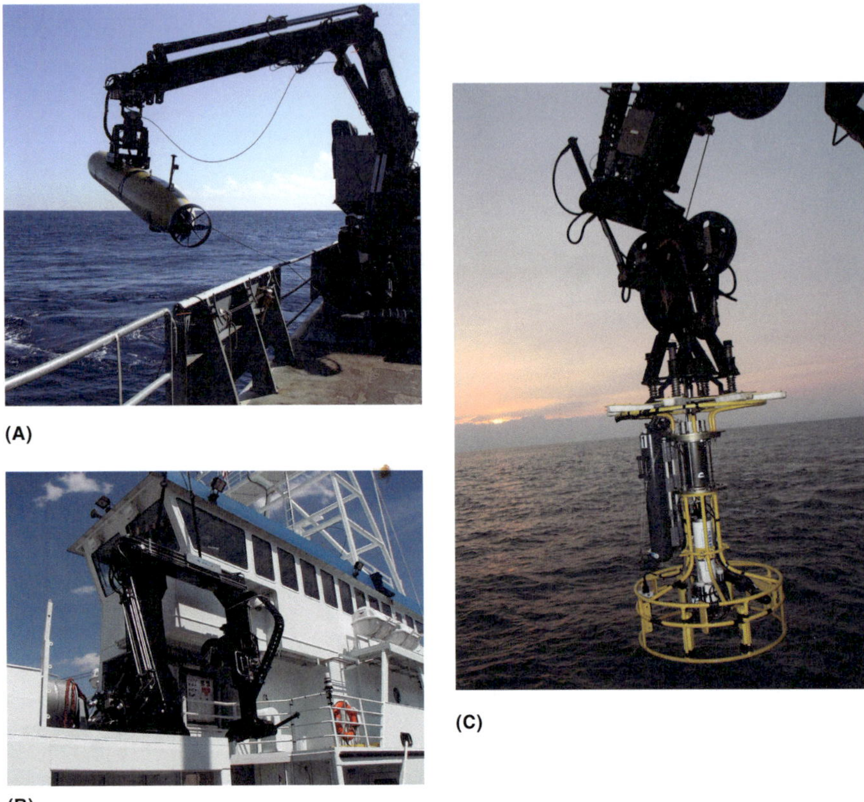

FIGURE 4-1 (A) An AUV being deployed using a custom OTS handling system (used with permission from ODIM Brooke Ocean). (B) The hands-free CTD handling system mounted on the *R/V Sharp*, which allows the CTD to be deployed and recovered without personnel holding the rosette. (C) A CTD deployed using the *R/V Sharp*'s OTS CTD handling system. The motion compensating function keeps the CTD at designated depth without regard to the motion of the ship, once deployed. (B and C used with permission from William Byam, University of Delaware).

operations are likely to exacerbate this issue. Existing options, including using a small boat or a grapple to hook gliders, AUVs, or ROVs, will be less viable in rough weather conditions. Development of over-the-side (OTS) lifting equipment, either portable or permanent, will be necessary to protect equipment and personnel. However, designing handling equipment that is optimized for current OTS equipment could negatively impact vessel utility over the 30-year lifespan of a ship. Instead, this type of equipment should be designed with future needs in mind.

Acoustic Quieting

Acoustic quieting requirements are essential for many missions (e.g., shipborne acoustic sensors, acoustic releases on equipment, offboard platforms with acoustic communications). Double raft mounting and/or resilient mounting will be increasingly desirable. Achieving compliance with ship-radiated noise recommendations set forth in the International Council for the Exploration of the Sea (ICES) report *Underwater Noise of Research Vessels* (commonly referred to as ICES 209; Mitson, 1995) is likely to be costly, and mission needs must clearly warrant imposition of this requirement if costs are to be minimized. Some recent and planned vessels, including the ARRV and RRS *Discovery*, are attempting partial compliance with ICES 209 specifications for a manageable and economic solution to ship-radiated noise.

Attention should also be paid to ambient noise and its impacts on habitability for the ship crew and science party, especially when round-the-clock operations are undertaken. The positioning of berthing and accommodations should be designed to avoid unnecessary and disturbing ambient noise.

Dynamic Positioning

Dynamic positioning is critical to handle deployment, recovery, and operation of offboard vehicles safely. Design conditions should strive to maintain position beam-on in at least sea state 6-7, 30-knot winds gusting to 40 knots, and a 0.5-knot surface current all from the same direction (Williams and Hawkins, 2009). The current Ocean class Science Mission Requirements (SMR) require that the ship be designed to maintain position in sea state 5, a 35-knot wind, and a 2-knot current (UNOLS Fleet Improvement Committee, 2003b).

Laboratories and Working Decks

There will be a continued need for plentiful laboratory and working deck space and capabilities. Laboratory space should be divided between ultraclean, clean, normal, and temperature-controlled areas, with sufficient flexibility to be used for multiple needs (Williams and Hawkins, 2009). There should be ease of and logical access into and between lab spaces for personnel and sample movements. Vessel design should include a substantial scientific stores area, including areas for frozen and refrigerated sample storage (Daidola, 2004).

Working deck design must be open and clear, with tie-downs for equipment and containers. There should be flexible deck space to support the use of laboratory and equipment vans, and easy and safe access to covered working areas using integrated overhead lifting gear. Decks must be able to handle increasingly heavy gear, including moorings, fleets of autonomous vehicles, and ROV equipment and winches. Freeboard should be as low as possible to allow for optimal handling of over-the-side equipment while keeping decks dry.

Berthing and Accommodations

Accommodation trends aboard research vessels include more single berthing for crew, specialized technicians, and scientists; berthing with natural light to promote natural sleep patterns; and galley and relaxation spaces that promote a healthy lifestyle at sea (Williams and Hawkins, 2009). The quality and design of crew living spaces are paramount for employee retention and morale. Specifications for noise levels and environmental conditions in both interior laboratory spaces and living quarters should strive to minimize ambient noise levels.

Other Design Attributes

A number of other scientific and operational trends will drive oceanographic ship design in the future (Daidola, 2004; Williams and Hawkins, 2009). These include the following:

- Larger, multidisciplinary science parties to make the best use of the ship resources and collect interdisciplinary and/or complementary data
- Longer cruise durations ranging over larger areas of the ocean
- Increasing desire to work in areas of rougher weather, demanding vessels capable of operating in higher sea states
- Specifications that comply with the Americans with Disabilities Act (ADA)
- 24/7 operations
- Higher-resolution and specialized hull-mounted swath bathymetry and sonar systems
- Larger and heavier pieces of portable science equipment
- Deployment, recovery, and maintenance of specialized offboard equipment
- More specialists (in addition to marine technicians) to service complex equipment
- Operational safety

The impact of these trends on dimensions and displacement is discussed later in this chapter.

DESIGN CHARACTERISTICS AND DESIGN DRIVERS

Table 4-2 displays ship design characteristics that are dictated by science needs as well as other characteristics inherent to setting future mission requirements that may have a significant cost impact. These design drivers are assessed by their priority (1-9, with 9 being the highest), established by the scientific community, and by their degree of ship impact (low-high), assessed by naval architects (UNOLS Fleet Improvement Committee, 2003b; Dan Rolland, personal communication, 2009). A "high" impact means that the ship's capital cost will increase if that requirement is met. For example, dynamic positioning is important for many types of science missions and has a large impact on ship design. The thrust delivery and control required add significantly to the ship construction cost, but given the high associated priority, dynamic positioning is likely to be an investment with widespread use. Conversely, aiming for higher ship speeds also has strong impacts on ship construction cost, but with a much lower priority. This indicates that when ship mission requirements are set, care should be taken to fully justify any speed that is on the steep side of the power curve. A corollary impact of higher speed is greater fuel consumption, leading to increased operating cost, and greater fuel tank volume, which can increase ship cost.

Efficiency

Efficiency is a vital consideration in the design of future oceanographic ships. Seeking a design with high propulsion efficiencies will lead not only to a lower operating cost but to a "greener" ship. Efforts to be more environmentally friendly often result in the addition of equipment to reduce emissions, which requires space in and adds weight to the ship in addition to its own costs, increasing ship construction costs. However, the potential for stronger regulations on emissions in particular local or regional areas (exist in the North Sea Sulfur Oxide Emission Control Area; International Maritime Organization, 1997) will affect ship design requirements and will not be achievable with current UNOLS vessels. Future oceanographic ship design may have to anticipate this by creating space and weight to comply with as-yet-undefined requirements or by accepting construction and operation cost increases associated with emission reduction measures. Other control measures, such as a carbon tax, could also drastically change the economics of traditional propulsion plants.

Recent increases in fuel costs dictate that high priority should be given to improving propulsion plant efficiency and reducing ship hull resistance. Many recent academic research vessels, such as *Atlantis* and *Kilo Moana*, have used some form of electric propulsion, and currently the Navy is contemplating shifting its combatant fleet toward integrated

Table 4-2 Research Vessel Design Drivers

Ship Design Driver	Priority	Ship Impact
ABS class/USCG certified	9	High
ADA accessibility	9	High
Working deck area and arrangement	9	High
Laboratory area and arrangement	9	High
Draft (less than 20 feet)	9	Moderate
Dynamic positioning capability	9	High
Fuel efficiency	9	Moderate
Maneuverability at slow speeds	9	Moderate
Sonar self noise	9	High
Bubble sweepdown	9	High
Seakeeping	8	High
Number of science accommodations	8	High
Crane handling on deck and on/off ship	8	High
Overboard handling operations	8	High
Overboard discharges/stack emission	8	Low
Other scientific echosounders	8	Moderate
AUV/ROV handling and servicing	7	Moderate
Workboat handling	7	Moderate
Science storage	7	Low
On deck incubations, locations/water	7	Low
Long coring capability	6	High
Mast location, met sensors	6	Moderate
Range[a]	6	High
Speed	6	High
Variable science payload	6	Moderate
Radiated noise[b]	6	High
One degree deep water multibeam	6	High
Endurance	5	Low
Ice strengthening	4	High
Marine mammal and bird observations	3	Low

[a]The committee thinks that "Range" deserves a higher priority than the value shown in this table, due to growing needs for ships capable of reaching distant research sites.

[b]The committee thinks that "Radiated noise" deserves a higher priority than shown on this table unless "Sonar self noise" (which has a high priority) is controlled.

SOURCE: Adapted from UNOLS Fleet Improvement Committee, 2003b; Dan Rolland, personal communication, 2009.

electric drives.[3] This trend has resulted in larger research and development expenditures for naval combatant electric propulsion, and future oceanographic ships are likely to benefit from advancements in power conditioning, reductions in plant size, and reductions in fuel consumption for a given power level.

There are other efficiencies to be considered. The performance of a research vessel is based upon the quantity and quality of the data it produces. A variety of issues can impact ship productivity, including the amount of time taken to deploy equipment to full depth and recover it, the time taken to change over from one piece of equipment to another, and time lost due to breakdowns in the winching and OTS handling equipment. This is increasingly important on multidisciplinary cruises, which often require capability for a variety of equipment to be used at any one site.

Although little can be done to improve deployment and recovery speeds through the water column due to the limiting hydrodynamics of the equipment and potential for damage due to overspeeding, the U.K. academic research vessel RRS *James Cook* was designed to substantially reduce the time for equipment changeover and breakdown losses. Winches are arranged to allow all wires to be permanently rigged up and quickly connected, while a system of sheaves allows any wire to be led over any of the main OTS handling equipment (Robin Williams, personal communication, 2009). These types of ship arrangements permit a high degree of integration and support diverse science objectives simultaneously, thus allowing more science to be carried out per day and increasing the ship's efficiency.

General Purpose and Specialized Design Requirements

Large general purpose vessels yield an economical long-term fleet that can satisfy uncertainty in future mission requirements. Although general purpose ships will serve a broad spectrum of future research activities, some scientific mission requirements will call for special purpose ships. These include fisheries surveying, which requires very quiet platforms; operations in the marginal ice zone, which result in specialized hull structure; deep submersible operations, which need strengthened A-frames and specialized hangar spaces; and three-dimensional (3D) seismic studies, which require large reinforced deck spaces to accommodate streamer reels, large-capacity compressors for air guns, rigging and booms for handling air gun arrays, and the ability to tow multiple air gun arrays and/or streamers (Daidola, 2004). Of these, seismic needs are currently

[3] For example, the Zumwalt-class destroyer DDG1000.

addressed with the *Marcus Langseth*; *Atlantis* serves as the tender for the *Alvin* manned submersible; and the NSF-funded ARRV will allow for work in marginal ice. These specialized ships are relatively young: *Marcus Langseth* was converted for research service in 2008, *Atlantis* was built in 1997, and the ARRV is anticipated to come online in 2014. Based on the evolving science and technology needs identified in Chapters 2 and 3 and the existence of capable specialized vessels, readily adaptable general purpose ship designs are most needed in the future fleet. The UNOLS fleet does not currently have any specialized fisheries vessels, although the National Oceanic and Atmospheric Administration (NOAA) operates four ultraquiet fisheries vessels and is slated to build three more by 2018 (Office of Marine and Aviation Operations, 2008; Tajr Hull, personal communication, 2009).

There are a number of ship design trends involving displacement and dimensions that are useful to consider, including (Williams and Hawkins, 2009)

- Increased beam, which increases damage survivability;
- Increased length, which improves the hull form for powering and control of bubble sweepdown over hull mounted transducers;
- Increased draft, which reduces bow emergence in a seaway and reduces bubble sweepdown; and
- Increased displacement, which supports increases in range, roll stabilization, science outfitting, and over-the-side lifting equipment weights.

Beam has been increasing as a result of stronger standards for damage stability but is likely to stabilize. Draft has also increased over time, likely due to the need to minimize bubble sweepdown for hull-mounted sonar systems. Minimization of bubble sweepdown has proven to be extremely challenging and can be a significant design driver for ships carrying these devices (Robin Williams, personal communication, 2009). Increasing beam and draft for conventional hull forms implies increased displacement, which leads to higher costs for ship construction. However, larger ships capable of carrying more scientists and performing more scientific experiments do provide an economy of scale. While adding more berthing and lab space increases ship construction costs, the cost per scientist decreases. This is supported by UNOLS statistics from 2008, where the average daily cost per scientist was higher for the Ocean ($1,062) and Intermediate ($982) classes than for the Global class ($946; data from UNOLS office, 2009).

International Maritime Organization (IMO) MARPOL Regulations

The United States is a party to Annex 1 of the IMO's International Convention for the Prevention of Pollution from Ships (MARPOL), which regulates oil pollution.[4] A 2007 amendment to Annex 1 is likely to have a significant effect on the design, cost, and operation of future research vessels. Ships with fuel capacity of more than 600 m^3 will be required to enclose the fuel tanks within a double hull. Several of the current Global class vessels (*Revelle, Atlantis, Thompson,* and *Langseth*) have fuel tanks with greater capacity.

This regulation has the potential to severely restrict the range of larger ships of the academic fleet, which in turn will affect scientific activities. Although ships built using Navy funds could be exempt from these regulations, the amendment provides a significant driver toward more fuel-efficient operations, including lower transit speeds, more streamlined hull forms, and efficient power generation and distribution systems for future Global and Ocean class vessels.

THE SHIP ACQUISTION PROCESS

The Navy's acquisition process related to the academic fleet has a significant impact on both ship cost and quality. The time from concept to delivery of any ship constructed with federal funds is extraordinarily long: the proposed new polar icebreaker is projected to take 8 to 10 years to enter service (National Research Council, 2007), and the new ARRV has taken more than 30 years of planning (*http://www.sfos.uaf.edu/arrv/*). Because of the lead times involved, it is vital that the most capable ship is constructed. Since decisions made at the earliest stage of design can have the greatest impact on the life-cycle cost of a ship (Bole and Forrest, 2005), science users need to participate in setting initial requirements and design specifications and to be included in the evolution of the design. This is especially important when the research requirements are translated into ship specifications, because poor decisions at this stage often yield a ship that will be unsatisfactory or uneconomical to operate.

One strategy that almost guarantees an unsatisfactory solution is the use of poorly defined performance specifications. Shipbuilding is a business, and shipbuilders must compete for contracts that are usually awarded to the lowest bidder. If specifications are not tightly defined, the shipbuilder may use inexpensive and unsatisfactory approaches to construction. Some of the recent UNOLS vessels procured through the Navy acquisition process have been constructed with poor attention to

[4] *http://www.imo.org/Conventions/contents.asp?doc_id=678&topic_id=258#7.*

detail because of this approach. Examples include the use of iron piping instead of copper-nickel for potable water systems because pipe material was not defined (as on *Thompson*), or deck drains that are not located at the local low point (thereby not working effectively) because the designer failed to specify a location (on *Atlantis*). There have even been cases where the drain piping has been run against grade (both *Revelle* and *Atlantis*). There is simply no substitute for specificity in fixed-price contracts, such as those the Navy uses to procure academic ships.

While cost constraints may preclude securing a ship with every desired specification, improvements could be made to the current system. Since hull structure is one of the cheapest aspects of a complete ship, one alternative to the current approach might be to consider building a larger ship than may appear to be affordable and bid certain scientific systems separately. This would allow for "mix-and-matching" the systems, creating a ship that does some part of the overall mission very well. Other capabilities could be deferred for a future refit, with unfinished space left for future equipment purchases and installation. Another alternative would be for the procuring agency to purchase certain high-tech equipment separately and provide it to the shipbuilder for installation, ensuring that the desired equipment is installed rather than a lower-cost component that would require replacement and increase life-cycle costs. One caveat with this approach is that equipment must be delivered to the shipyard on time, and any required interfaces with the ship must be correctly and precisely defined. If this is not done, the shipyard will likely consume all potential cost savings by claiming increased costs due to delay and disruption associated with failure to be timely and properly defined. A common hull design between vessels of each class, as done previously with Global class ships (i.e., *Thompson*, *Atlantis*, *Revelle*, and the NOAA ship *Ronald H. Brown*), could also provide cost savings.

NSF created a design and construction plan for the AARV that was intended to address many of the problems that have impacted earlier oceanographic ship acquisition programs. The ARRV process involves the scientific user community in the design and construction of an oceanographic ship from the preconstruction phase through post delivery of the ship. It is summarized in Box 4-1.

CONCLUSIONS

The fleet of the future will be required to support increasingly complex, multidisciplinary, multi-investigator research. The design of future oceanographic ships is likely to become more challenging in order to achieve the needed integration and balance of facilities and equipment. Multidisciplinary, multi-investigator cruises will drive many aspects of

BOX 4-1
The ARRV Procurement Process

The ARRV is being built under the direction of NSF to support research in coastal and open ocean settings, particularly in those regions that experience moderate seasonal ice. ARRV, as the first ice-strengthened ship to join the academic fleet, requires special capabilities and presented engineering challenges that do not apply to more general purpose vessels. In order to provide strict oversight for vessel fabrication, NSF implemented a four-phase building project that required successful completion of early phases before funding would be awarded for subsequent phases. The phases included a project refresh (design review), yard selection and acquisition, ship construction, and delivery and transitions to operations.

A key element of the process was the creation of an ARRV Oversight Committee to obtain community input and advice on ship design and construction during all of the phases. This included a review of a final refreshed design and de-scoping plan, draft shipyard contract, and shipyard scope of work; a periodic review of ARRV construction progress; review of delivery voyage and the shakedown science test cruises; and review of warranty period and final acceptance.

The oversight committee provides advice on the establishment of design and budget priorities, ensuring that construction remains within the agreed scope and cost. The committee was established and supported by the University of Alaska, Fairbanks (UAF), and its membership and scope of activities are approved by NSF. The committee is responsive to NSF and UAF by providing reports that detail and track the status of recommendations. The committee's membership is fluid and may change depending on needed expertise for each phase of design, construction and trials.

The ARRV procurement process entails a competitive two-step shipyard selection process. Step 1 is the competitive qualification of shipyards through a technical proposal submission. Step 2 is a best-value price competition among acceptable shipyards in response to a request for cost proposals. Shipyards that do not pass Step 1 are expected to be eliminated to reduce risks of procurement delay, allow fewer potential protest risks or expenses, and maintain strong price competition among acceptable shipyards. The shipyard selection process begins with a request that interested shipyards demonstrate their qualifications for the ARRV project. The request includes the baseline project design package, a thorough description of the selection process (including evaluation methods), and detailed instructions to the potential offerors.

design, including power plant and propulsion, laboratory and working deck layout, over-the-side handling, launch and recovery, and equipment changeover. Larger science parties and more complex technology will require more laboratory and berthing space. The growing trend toward use of multiple offboard vehicles will also impact the design with respect to freeboard and deck space. Vessel design will have to incorporate technology that is currently available, such as dynamic positioning or state-of-the-art sonar, while remaining adaptable for future technological upgrades. The capability to operate in high latitudes and high sea states will also be required.

Because technology changes rapidly and ship lifespans are long, future academic vessel designs need to be general purpose and highly adaptable to changing science needs. Specialized ships will also be needed for some disciplines, with designs that are well matched to disciplinary needs while also being available for limited general purpose work. Trends toward increasing beam, length, draft, and displacement and the economy of scale present in larger hulls suggest that investments in larger, more capable vessels in any size class are preferred.

The current Navy ship acquisition process does not emphasize inclusion of the scientific community in decision making regarding academic ship design and specifications. Development of the NSF-sponsored ARRV has benefited from community-driven ship design, allowing the users to participate more fully and create optimal designs for the cost constraints.

5

Ship Time Costs and Their Impacts

How the increasing cost of ship time will affect the types of science done aboard ships.

One of the most serious issues facing federal agencies that support shipborne science, ship operating institutions, and science at sea itself is the increasing cost of operating research vessels. Higher ship costs will almost certainly force significant changes in the way U.S. academic research ships are scheduled and used. This issue has been studied in recent years by committees convened by the University-National Oceanographic Laboratory System (UNOLS), federal agencies and their advisory boards, and independent commissions (e.g., U.S. Commission on Ocean Policy, 2004; Betzer et al., 2005; McNutt et al., 2005; Collins et al., 2006; UNOLS Fleet Improvement Committee, 2009). Trends relating to the cost of ship time for the UNOLS fleet are examined in this chapter. These include major cost factors and trends, the relationship of research ship scheduling to operational costs, the potential for expeditionary planning, future trends in fleet composition, and ship layups. The impacts of these cost trends are also examined in the context of their effect on research proposals and awards, the efficiency of ship operations, and their potential to alter the present operating model.

SHIP TIME COST TRENDS

The primary expenses of research ship operation are crew costs, fuel costs, maintenance and overhaul, technical and shore support, and consumables. These costs for the UNOLS fleet between 2000 and 2008 are shown in Figure 5-1. Crew and fuel costs are the two largest single components of total research vessel operating costs, accounting for approximately 50 percent of total operating costs in this period, although the impact of fuel costs on total costs more than doubled over nine years. While "all other costs" also appears to be a significant factor, it is driven by fleet indirect costs, which are proportional to direct costs. Indirect costs make up between 37 and 46 percent of the category's costs, while the rest of the category (food, insurance, equipment and supplies, travel, shore facility support, and miscellaneous costs) has individual costs of approximately $3 million or less per year.

The increase in overall UNOLS fleet costs from 2000 to 2008 was not

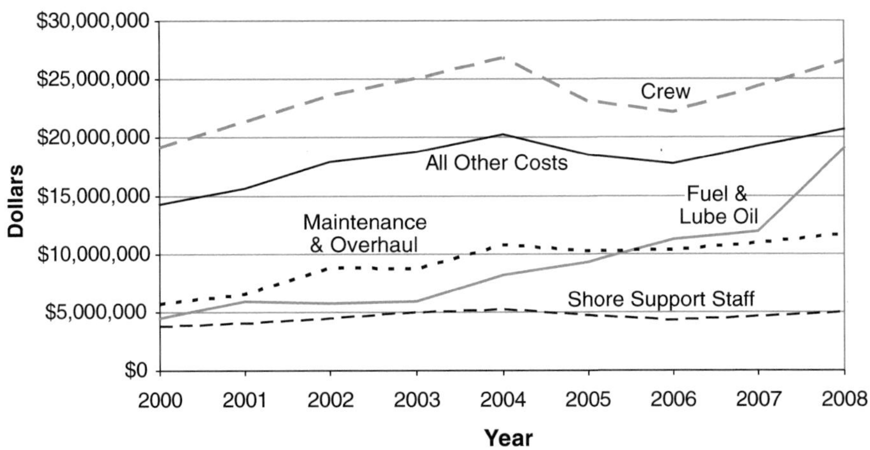

FIGURE 5-1 Major cost factors for the UNOLS fleet, 2000-2008. The categories listed are crew salaries and benefits (dashed gray line), fuel and lube oil (solid gray line), maintenance and overhaul (small dashed black line), shore support staff (long dashed black line), and all other costs (solid black line). The category of all other costs includes food, insurance, equipment and supplies, travel, shore facility support, indirect costs, and miscellaneous costs. This figure includes both estimated and actual costs from ship proposals. In several cases, total operating costs for individual ships are missing. This is most often the case with Local class vessels, and it is not expected to significantly impact the total costs. The 2008 costs associated with *Marcus Langseth* are not included (data from the UNOLS Office, 2009).

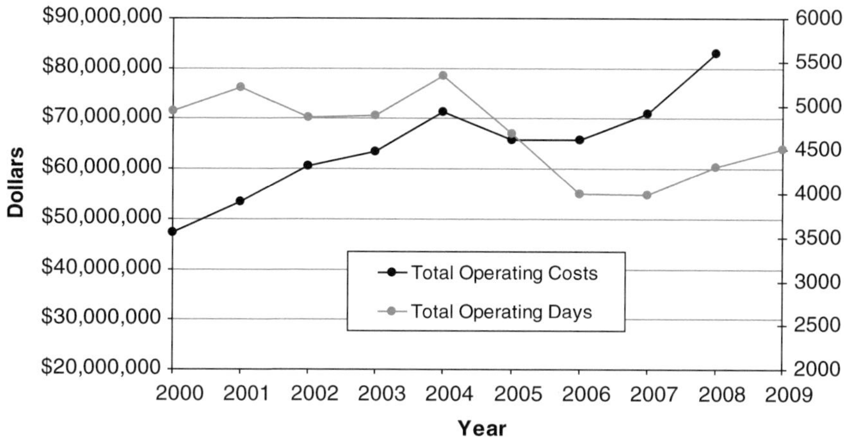

FIGURE 5-2 UNOLS fleet total operating costs (black) versus number of ship days (gray) (data from the UNOLS Office, 2009).

due to an increase in the total number of operating days for the fleet. In fact, ship operating days declined 13 percent from 2000 to 2008, while total costs increased 75 percent (Figure 5-2), meaning that the average cost per ship day doubled in that same period. Should long-term ship costs continue to increase at rates comparable to those of the last decade, it is very likely to pose severe problems for research ship operators and federal funding agencies alike.

Crew Costs

Crew salaries and benefits are consistently among the greatest cost drivers for the academic fleet (an average of $23.3 million per year from 2000 to 2008; data from the UNOLS office, 2009). Crew sizes on Ocean and Global class vessels are regulated by the U.S. Coast Guard (46 CFR §188-196), so there is little room for cost savings through personnel reductions. In addition, UNOLS vessels must comply with new environmental, safety, and security regulations. These new measures have required increased crew training and increased staffing requirements (UNOLS Fleet Improvement Committee, 2009). Operating institutions also face salary competition, especially for marine engineers, from other industries (i.e., cruise ships, offshore oil and gas).

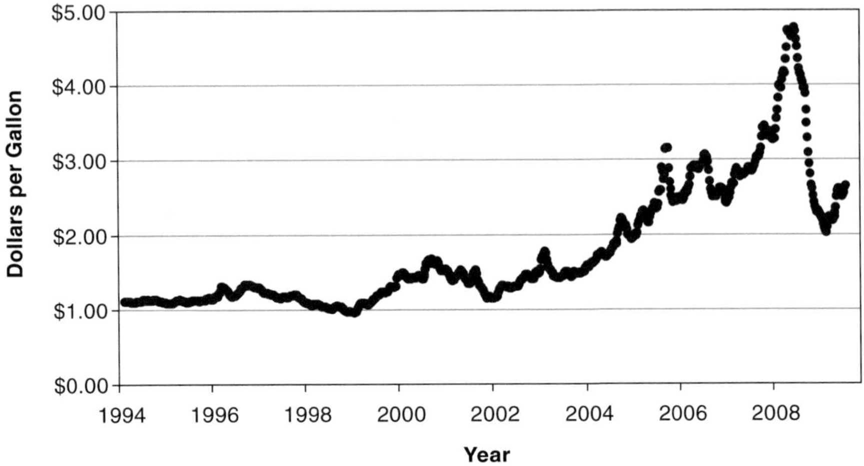

FIGURE 5-3 U.S. retail diesel prices, 1994-2009 (data from the Energy Information Administration, August 2009).

Fuel Costs

Fuel consumption is proportional to the size and speed of the vessel. Research vessels, which are equipped with diesel electric drives and typically operated at low speeds, do not lend themselves to significant future efficiencies. The type of research done aboard ship can also affect fuel costs on individual programs. For example, a surveying cruise that maintains a sustained ship speed will use more fuel than a research cruise with short transits between stations.

Due to recent market volatility in the price of crude oil (Figure 5-3), fuel costs for the fleet have escalated over the past few years. Between 2005 and 2008, fuel expenses doubled from $9.3 million to $18.6 million per year (data from the UNOLS office, 2009). In 2009, fuel prices dropped substantially from their 2008 highs, but the unstable nature of recent oil prices suggests that fuel costs may remain a significant aspect of total operating costs in the future. However, future market controls or carbon emission legislation could have significant and as yet unknown impacts on the price and rate increases of fuel.

Ship Schedules and Fleet Management

Research ship schedules are managed through UNOLS with the goal of federal oversight specifically to seek maximum efficiency. Ship day rates (the metric used by supporting federal agencies) are influenced by scheduling because fixed operating costs affecting the daily rate are not proportional to the number of days each vessel is used each year (e.g., costs of full-time crew, scheduled maintenance, shore support, regulatory compliance). Full schedules (referred to as "efficient" schedules) lower the daily at-sea rate by combining these fixed costs with incremental operational costs over a greater number of days.

The present ship scheduling process produces a one-calendar-year schedule, beginning six to eight months ahead. There is an attempt to build full schedules for each ship from an ad hoc collection of cruises that federal agencies indicate are likely to be funded. However, it is nearly universal for ship schedules to change based on the differing funding and decision time scales among supporting agencies, often with late notice compared to the National Science Foundation (NSF) funding time frame (Rose Dufour, personal communication, 2009). This leads to schedules that are not finalized until after the beginning of the operating year, often with some remaining uncertainty. In addition, some flexibility is built into schedules to account for episodic events of scientific interest (e.g., volcanic eruptions, harmful algal blooms). Similarly, flexibility is needed for cruises that are rescheduled or canceled on short notice due to lost or damaged equipment or societal events (e.g., Indian Ocean piracy).

Recent years have seen some scheduling delays in an effort to prevent ships from being idled (discussed later in this chapter). Dividing scheduled projects among several ships of the same class aims to reduce unnecessary expenditures for partial layups, but also results in an increase in ship rate for the ships that have fewer days scheduled.

Operational costs are also impacted by the geographic distribution of the fleet, which does not mirror the requested locations for use. Regional class ships are more closely tied to the location of their operating institution than are larger vessels. The aging Intermediate class presents some scheduling problems because these ships tend to work closer to their home port. *Wecoma*'s Oregon location leads to a shorter operating season. The close proximity of *Endeavor* and *Oceanus* (Rhode Island and Cape Cod, respectively) to each other presents difficulty in creating full, more efficient schedules. This issue is exacerbated by the lack of capabilities on these ships, which makes them less desirable for research cruises.

Expeditionary Scheduling

One alternative to ad hoc annual scheduling is expeditionary scheduling. In this type of scheduling, an announcement is made that a ship will be operated in a specific region during a given time window. Proposals are then sought to use that opportunity. UNOLS does not consider any of its ships presently to be funded in expeditionary mode (Mike Prince, personal communication, 2009). With a few exceptions, noted below, expeditionary scheduling may not be desirable for the UNOLS fleet. Within each class, ships are somewhat interchangeable and it may not be advantageous to fix the operating region for future years.

The *Atlantis* (with *Alvin*) is generally tied to the annual window of opportunity on the Juan de Fuca Ridge and often spends the rest of the year near the East Pacific Rise. If the ship were scheduled in an expeditionary mode, a community workshop or panel could decide on other regions far enough in advance to allow for proposals to be submitted, reviewed, and awarded. At present very few proposals for *Alvin* are submitted or funded for work beyond the Juan de Fuca Ridge or East Pacific Rise. When such proposals are submitted, many state that either *Alvin* or *Jason* can be used, allowing schedulers to send ships equipped with *Jason* to these regions. As another example, the seismic vessel *Marcus Langseth* could possibly benefit from some expeditionary scheduling. Currently, the schedule is dominated by previously funded seismic work. Future years' schedules may require orderly movement from one area to the next to lower transit costs and time expended. As with *Atlantis*, a community workshop could provide recommendations for the most efficient use of the ship.

Fleet Composition and Science Impacts

The ship replacement and retirement plan outlined in the 2009 *UNOLS Fleet Improvement Plan* will reduce the academic research fleet by nearly 40 percent by 2025 (Figure 5-4; UNOLS Fleet Improvement Committee, 2009). The projected retirement of three Global class ships reduces overall ship sizes and could produce overall fleet economies. However, Global class vessels are presently the most heavily subscribed. Chapters 2 and 3 conclude that there will be increased demand for the large research vessels with their deck loading, berthing, and sea state capacities. The new and planned Ocean class ships are significantly less capable than the Global class in terms of deck loads and berthing. Accommodating heavy deck loads and large science parties on Ocean class vessels would require scheduling extra legs, leading to more time in port and a greater number of ship days per research mission. In addition, the current Ocean class ship, *Kilo Moana*, has a day rate that is comparable to the Global class (Figure 5-5). Thus, if day rates for the planned Ocean class vessels are similar to *Kilo Moana*, total operating costs for 2025 will not decrease. Furthermore, with the planned addition of Ocean class vessels, there will be fewer vessels able to support the widest-ranging, most resource-intensive marine science research programs of the future and the decrease

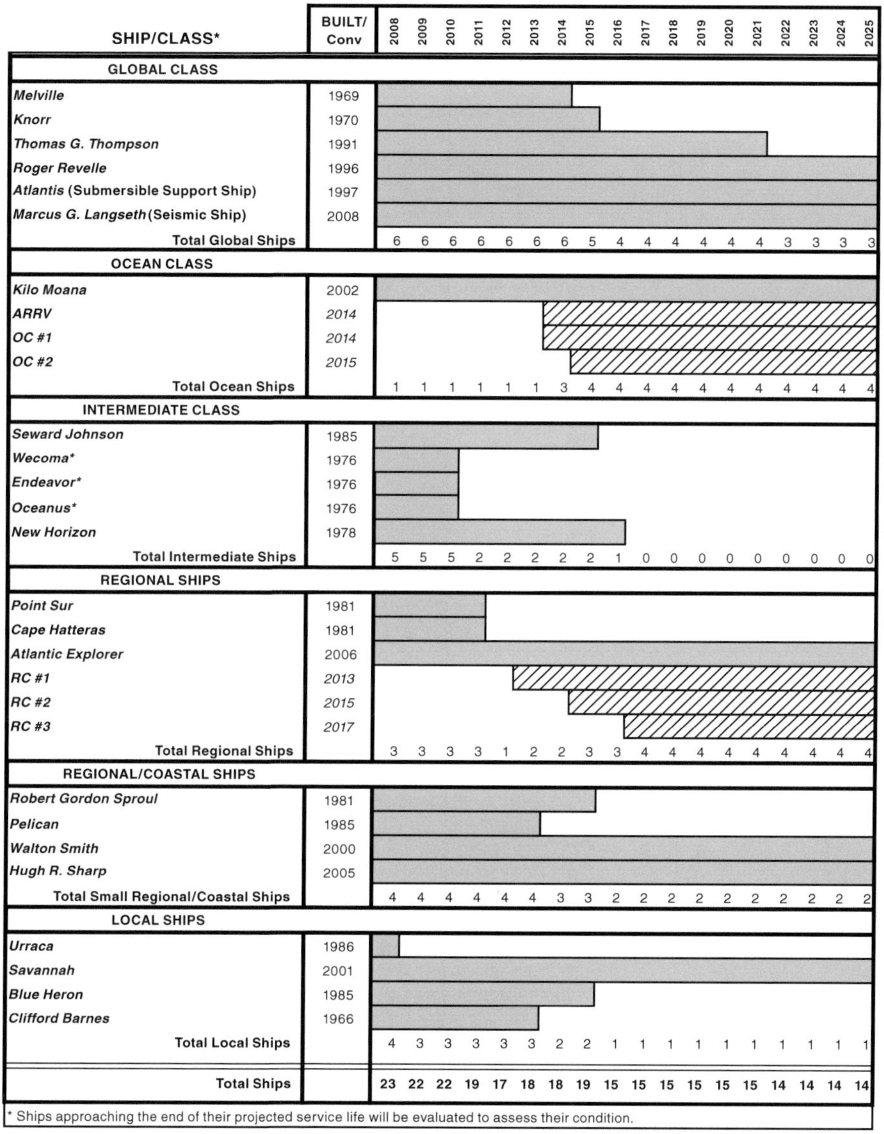

FIGURE 5-4 The UNOLS fleet projected service life time line (adapted from UNOLS Fleet Improvement Committee, 2009; used with permission from UNOLS).

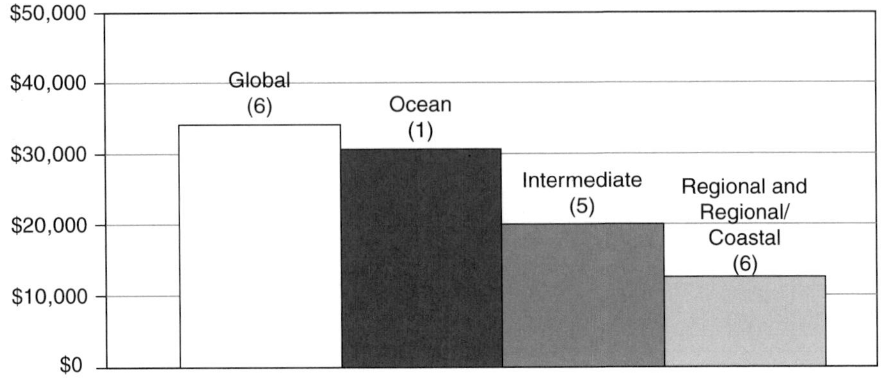

FIGURE 5-5 The average daily operating rate by UNOLS class, 2008. The number of ships in each class is in parentheses after the class name. For example, the Intermediate class has five ships (data from the UNOLS Office, 2009).

in overall fleet size will create greater difficulty in scheduling multiship operations.

An additional consideration is the cost per science berth. Global vessels, although the most expensive ships of the fleet with the highest day rates, are in high demand and have heavy usage (UNOLS Fleet Improvement Committee, 2009). Thus, as discussed in Chapter 4, the Global class ships are less expensive than the Ocean and Intermediate classes when their day rate is divided by the number of science berths aboard. This economy of scale suggests that there are advantages to building larger ships with more science berths and more deck and payload space.

Scheduling General Purpose Ships with Specialized Facilities

The specialized ships (*Atlantis* and *Langseth*) create their own scheduling niches because they fulfill research missions that cannot be accommodated by general purpose vessels. The specialized facilities on some of the general purpose Global class vessels (i.e., long coring facility on *Knorr*, sonar system on *Revelle*) also attract science missions. Increasing scheduling efficiency for the general purpose vessels may lead to difficulty in scheduling cruises that require specialized facilities. Some redundancy may have to be built into these ship schedules.

Idle Periods And Layups

All research ships have idle periods when they are not carrying out work at sea. Required maintenance, training, and inspection activities are part of this idle time and are included in the ship's normal schedule. Factors that affect idle periods on research ships include planned major maintenance periods, variations in funding levels for major field programs, uncertainties in federal research budgets, inconsistent business relationships with nonfederal users, and seasonal demands. Some excess ship capacity is occasionally used in each class to handle planned maintenance periods, which require fairly long downtimes. Global class ships carry larger crews, have a wider variety of maintenance and repair equipment, and often have longer transits, so there is opportunity to carry out more routine maintenance at sea than on smaller vessels.

Large field programs, such as the Joint Global Ocean Flux Study (JGOFS), Climate Variability and Predictability (CLIVAR), and Global Ocean Ecosystem Dynamics (GLOBEC), also affect idle periods (Mike Prince, personal communication, 2009). These types of programs require ships to be in specific locations at specific times and can also require multiship operations, increasing the difficulty of creating efficient schedules for other funded programs. At-sea support required for major programs in the future, such as the Ocean Observatories Initiative (OOI), may place even heavier demands on the fleet in certain locations in the future.

Federal budget uncertainties in recent years have made efficient ship scheduling and full utilization even more difficult. Some federal agencies that support ship use are forced to withdraw their already scheduled requests or do not place ship time requests until after annual research ship schedules are determined. Nonfederal users can help strengthen ship schedules, but are subject to restrictions including Navy approval. Ship schedulers tend to seek out these partners only when openings appear in schedules, creating an inconsistent and unsatisfactory business relationship. Additionally, outside users are not held to a contract, which allows them to withdraw at late dates and create further holes in ship schedules.

Seasonal demand also affects lay-ups. Peak ship demand during late spring to fall time frame normally comes close to fully utilizing the fleet (Figure 5-6), but winter demands are reduced especially for those ships operating in areas with rougher winter weather.

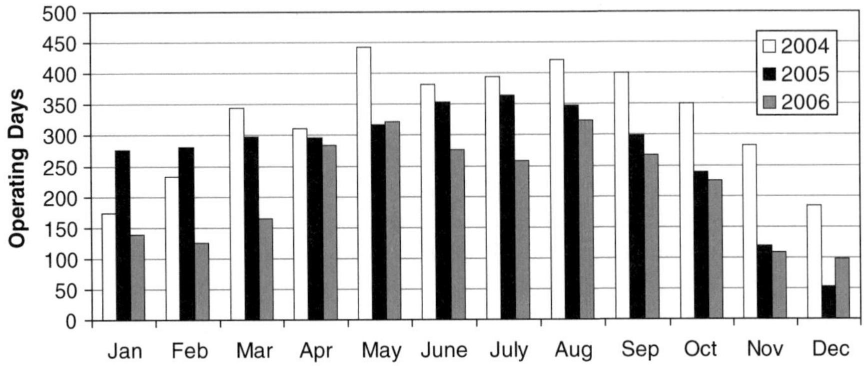

FIGURE 5-6 The UNOLS fleet seasonal utilization trends from 2004 to 2006 (adapted from UNOLS Fleet Improvement Committee, 2009; used with permission from UNOLS).

THE IMPACTS OF INCREASING SHIP COSTS

Agency Proposals Requiring Ship Support

The most direct impact of rising ship costs on science has been a decline in the total number of funded ship days from 2004 onward (Figure 5-2). NSF's annual budgets, including those for divisions that support research requiring the oceanographic fleet, have been funded at levels below the inflation rate. This is a serious mismatch to ship costs, which at the same time have risen at more than three times the inflation rate. Despite the rising number of days at sea requested in proposals (Figure 5-7), NSF could not maintain its non-ship-related research without reductions in the number of ship days funded each year (UNOLS Fleet Improvement Committee, 2009). An initial agency response was to defer ship time into future years to meet shrinking federal budget levels, with the hope that future budget increases would provide relief. In 2005, more than 500 ship days were deferred to future years.

Proposal success rates also factor into ship demand. The success rate of proposals in the NSF Ocean Sciences Division (OCE) is higher than the overall NSF success rate (Figure 5-8). However, the rate of successful proposals with ship time has declined, while the overall proposal success rate within OCE has increased. For example, 33 percent of 1349 proposals submitted to OCE in 2008 were funded, whereas only 21 percent of 255 proposals with ship days were successful. This leaves an obvious imbalance in the number of ship days funded versus the number of ship days requested, as shown in Figure 5-7. While some OCE proposals related to the UNOLS fleet but not including ship time (including those that sup-

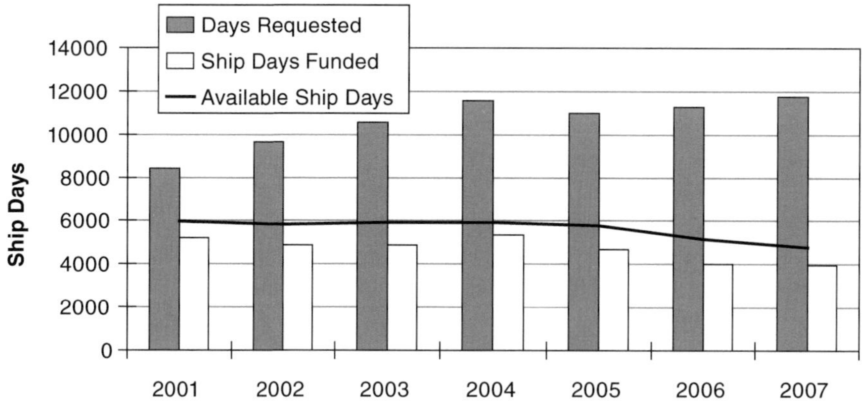

FIGURE 5-7 UNOLS ship time demand versus days funded, 2001 to 2007. The number of days requested is shown in gray, the number of funded ship days is shown in white, and the number of available ship days is a black line (adapted from UNOLS Fleet Improvement Committee, 2009; used with permission from UNOLS).

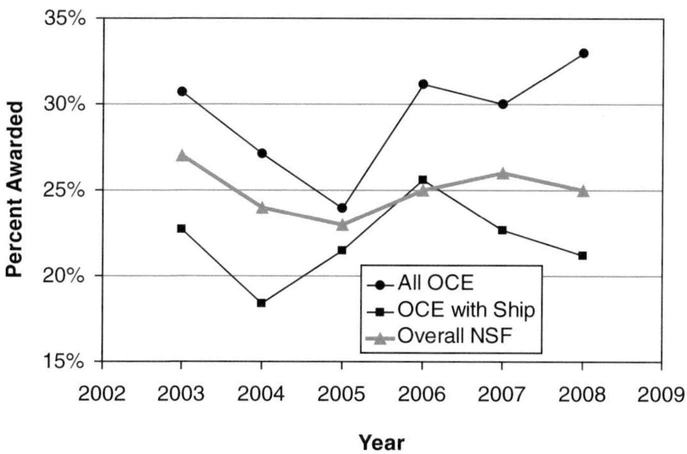

FIGURE 5-8 The success rate of proposals submitted to the National Science Foundation (NSF) overall, to the Ocean Sciences Division (OCE), and to OCE that have requested ship use (data from NSF, 2009).

port ship operations, marine technicians, and other facilities) have very high success rates, the low rate of successful proposals with ship time is detrimental to maintaining the UNOLS fleet.

Efficiency of Ship Operations

One common metric of efficiency in ship operation is the fullness of each ship schedule. Projected reductions in fleet capacity due to vessel retirements implies that if the present level of ship days funded per year is maintained, the fleet would be fully supported by approximately 2011 (Figure 5-9). A more difficult measure of efficiency is the productivity of each ship day. Anecdotal accounts suggest that funding agencies and the research community are more effectively linking and combining programs to make the best use of ship time. The rise of multidisciplinary, multi-investigator programs discussed in Chapter 2 implies continued progress in this measure of efficiency.

It is worth noting that greater efficiency in operations and scheduling does not imply cost decreases. Working efficiently may actually mean working more expensively. For example, costs per day could escalate quickly on a ship that is simultaneously operating several AUVs while also involved in other types of sampling or data collection. More deck, bridge, and technical support personnel may be required to assist with the various operations, contributing to a higher daily operating rate. In addi-

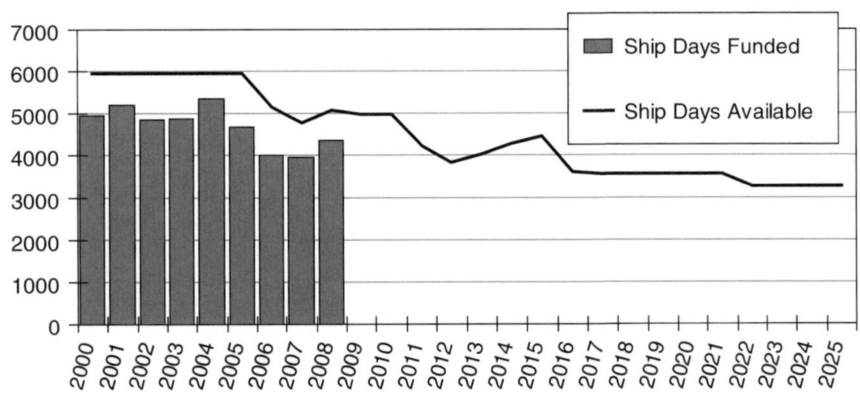

FIGURE 5-9 The difference between UNOLS funded ship days and the ship days available for use, 2000-2025. Funded ship days are shown in gray, while available ship days are shown with a black line. This graph assumes that all ships are capable of being fully utilized to support science operations. Realistically, some of the older and less capable vessels are likely to continue to be undersubscribed (adapted from UNOLS Fleet Improvement Committee, 2009; used with permission from UNOLS).

tion, multi-investigator cruises may lead to more idle time for scientists onboard, increasing science team costs.

Alterations to the Present UNOLS Model

The present model of UNOLS funding and operation is sensitive to the robustness of fiscal support. If future federal budgets supporting ocean science increase at the same rate as ship operation costs, the community can retain the flexible scheduling and excess capacity that has worked in the past. If not, UNOLS and the ocean research community will be required to use existing resources more effectively (Mike Prince, personal communication, 2009). Because the scheduling process already attempts to maximize efficiency, further costs savings may require a longer time horizon for planning. Although budget issues in the past few years have caused ship scheduling to occur later in the year, there is growing pressure to schedule the Global class ships significantly earlier than is the current practice.

Another trend may be toward increasing the flexibility of cruise timing, especially if it requires specialized equipment supported by only one ship or involves work in remote locations. For these areas, already-funded research projects may be deferred to a later year when there is enough demand. This can cause difficulties for research programs that require repeat surveys, recovery of deployed instruments, or significant international cooperation.

A third option is to average the fleet costs over a multiyear period. This would stabilize the day rate for a certain amount of time and increase the possibility of working with nontraditional funding sources. Finally, it may be worthwhile to investigate the possibility of home-porting ships in common locations to take advantage of cost savings and provide geographic proximity to research areas.

CONCLUSIONS

Due to insufficient funds to support research on increasingly expensive ships, the number of ship days requested is rapidly outpacing operational days. Crew and fuel costs are likely to continue as significant factors in total operational fleet costs. The push for more efficient ship scheduling may lead to longer lead times for research projects and reductions in the ability of the future fleet to accommodate late-breaking scientific and funding opportunities. Present trends in science and technology indicate further growth in major research programs requiring significant ship resources. The increasing cost of ship time and economies of scale may lead to greater use of Global class UNOLS vessels, which are capable of simultaneously carrying out multiple science operations. Complex programs are less likely to require multiple legs, lowering operational costs, if put on the largest ships of the fleet. The reliance on Ocean class vessels in the current fleet renewal strategy probably will not lead to a future fleet with reduced operational costs, but may lead to a fleet with fewer capabilities.

6

Partnerships

The usefulness of partnering mechanisms such as UNOLS to support national oceanographic research objectives.

The University-National Oceanographic Laboratory System (UNOLS) brings research scientists, ship operating institutions, and federal and state agencies together to coordinate economical and cost-effective use of the U.S. academic research fleet (see Box 1-1 for the UNOLS mission statement). As such, it provides a partnership mechanism to support national oceanographic research needs. This paradigm is related to other successful science partnerships; for example, the University Corporation for Atmospheric Research promotes understanding of the atmosphere through collaboration between federal agencies and academic institutions. This chapter reviews partnership goals and benefits, the Navy business model regarding UNOLS partnering, possibilities of increasing transparency and communication with other agencies and agency divisions, and international partnering facilitated by the UNOLS consortium.

THE PARTNERSHIP MECHANISM

UNOLS ships are operated as shared-use facilities that are equally available to a wide range of science community users. Having a number of operator institutions throughout the country promotes different perspectives, innovation, institutional and state support, and a certain amount of healthy competition (Bash, 2001). The UNOLS structure also promotes cooperation, through coordinated scheduling and the sharing of best practices from within and beyond the community (Mike Prince, personal communication, 2009). Tasking ocean science research institutions to operate research vessels ensures that the goals of the ship operators and research community are closely aligned.

PARTNERSHIP BENEFITS FOR PARTICIPATING FEDERAL AGENCIES

Federal agencies bring a variety of assets to the UNOLS consortium. The Navy owns all but one of the UNOLS Global and Ocean class vessels, while the National Science Foundation (NSF), academic institutions, and other research entities own the Regional and smaller research ships. Partnering with UNOLS permits federal agencies to do the following (Herr, 2006):

- Select the right size ship for each of their science and technology missions
- Share transit costs with other agencies and institutions
- Extend mission equipment outfitting to multiple agencies
- Allow multiple ship missions without chartering commercial vessels

The partnership allows each supporting federal agency to access the entire UNOLS fleet with its variety of ship sizes and capabilities, including the use of larger vessels for deep water and global research and smaller vessels for nearshore and coastal oceanography needs (Office of Naval Research, 2006).

The ability to "right size" a research vessel for specific missions can provide significant cost savings. A 2006 Naval Research Advisory Committee (NRAC) case study concluded that the Navy saved $46.3 million between 2001 and 2006 by utilizing the entire UNOLS fleet, rather than using only the Global and Ocean class vessels it owns (Herr, 2006). Other cost savings were realized by sharing transit and equipment costs with other agencies. Sharing transit, maintenance, and equipment costs with NSF through the UNOLS partnership structure reduced the Navy's costs by $11 million in the same time frame. Agreements between the Navy and NSF have resulted in significant NSF equipment expenditures for Navy-owned ships (including conductivity-temperature-depth sensors [CTDs], Acoustic Doppler Current Profilers [ADCPs], sonars, etc.). The use of UNOLS vessels for multiple ship missions, rather than chartered commercial ships, has also provided savings to the Navy (roughly 30 percent for a Global class size vessel). Leveraging through the UNOLS partnerships is estimated to have saved the Navy $76.2 million over 30 years (Herr, 2006; Office of Naval Research, 2006).

FUTURE PARTNERING OPPORTUNITIES WITHIN FEDERAL AGENCIES

NSF Office of Polar Programs

The Directorate for Geosciences (GEO) funds NSF's portion of the UNOLS consortium. However, NSF's Office of Polar Programs (OPP) collaborates with the U.S. Coast Guard (USCG) to operate the *Healy* in support of Arctic science programs. NSF, USCG, and UNOLS formed a subcommittee within UNOLS, the Arctic Icebreaker Coordinating Committee, to provide advice to the USCG in order to facilitate and enhance science aboard the icebreaker fleet.[1] OPP also supports research in the Antarctic using a variety of ships (including USCG ships *Polar Sea* and *Polar Star*, chartered vessels *Palmer* and *Gould*, the Russian ship *Krasin*, and the Swedish vessel *Oden*). These vessels serve multiple purposes, supporting oceanographic research and resupplying land-based research stations. The Antarctic vessels are operated by various contractors and agencies (including the USCG) and scheduled through OPP. The future Alaska Region Research Vessel (ARRV) will be a GEO-supported vessel scheduled through UNOLS. It is likely that OPP will be a major source of support for the ARRV.

The need for coordination between the high-latitude oceanographic research supported by NSF OPP and that supported by NSF GEO and other agencies is likely to increase as a result of the growing interest in high-latitude research requiring icebreaker or ice-strengthened capabilities. This includes research such as the role of sea ice loss in climate change and exploration of polar marine ecosystems, which are national ocean research priorities outlined in the *Ocean Research Priorities Plan* (Joint Subcommittee on Ocean Science and Technology, 2007). In recent years, there has been some progress toward integration of the polar vessels with the UNOLS fleet. Investigators can view OPP/USCG ship schedules and request these ship resources via the online UNOLS ship scheduling system. There are opportunities for further integration, such as sharing specialized seagoing technicians and instrumentation between the UNOLS and the OPP/USCG fleets.

---

[1] *http://www.unols.org/info/ucharter.html#annexVI*

NOAA

The National Oceanic and Atmospheric Administration's (NOAA's) participation in the UNOLS partnership differs significantly from that of NSF and the Navy. The NOAA-owned and operated Global class vessel *Ronald H. Brown*, although not a member of the UNOLS fleet, is scheduled in cooperation with UNOLS. NOAA also uses UNOLS ships to supplement their needs for required data collection. While NOAA's research fleet currently has a capacity of 4800 ship days per year, in 2008 the agency identified the need for a potential additional 13,200 days per year to fulfill its mission requirements (Office of Marine and Aviation Operations, 2008). From 2000 to 2008, NOAA used an average of 538 ship days per year on the UNOLS fleet (data from UNOLS Office, 2009). In those years, UNOLS vessels had an average of 768 ship days per year of unfunded capacity (often called excess capacity), due to agency funding limitations, equipment availability, and weather and time constraints (data from UNOLS Office, 2009; UNOLS Fleet Improvement Committee, 2009). Identifying future NOAA missions that can be carried out on ships of the UNOLS fleet would strengthen UNOLS ships schedules and reduce unfunded ship days, increasing the efficiency of the UNOLS fleet.

PARTICIPATION IN INTERNATIONAL PROGRAMS

International Research Ship Operators Meeting and International Marine Technicians Workshops

Many of the UNOLS federal agency partners participate in the International Research Ship Operators' Meeting (ISOM), an annual meeting that promotes discussions to improve services for research at sea, including ship time exchange between countries and updates on national research fleets. The exchange of knowledge, plans, and experience is especially important in an era of decreasing budgets but increasing societal relevance for coastal nations. ISOM also sponsors workshops and working groups, including the International Marine Technicians Workshops (INMARTECH), and ISOM members often participate in the UNOLS Research Vessel Operators Committee and Research Vessel Technical Enhancement Committee meetings.

INMARTECH facilitates an international exchange of knowledge and experience between marine technicians from both academe and industry. Specialists are invited to give presentations on selected subjects, and workshop sessions are arranged to foster informal discussions among the approximately 100 participants, including attendees from UNOLS ship operator institutions.

Ocean Facilities Exchange Group and Other Partners

The Ocean Facilities Exchange Group (OFEG) is a partnership of six European countries that barters research vessels and major equipment (*http://www.noc.soton.ac.uk/ofeg/pages/ofeg/index.php*). Much like UNOLS, the goal is to increase efficiency through bartered assets, shared transit costs, and a synchronized annual ship schedule. Although the United States actively barters ship time with several OFEG members (including the United Kingdom), further coordination of UNOLS and OFEG could increase the pool of available ships and increase scheduling efficiency. Additionally, UNOLS could more fully explore the possibilities of trading ship time with countries that are geographically distant (such as Korea, Japan, and India) to increase ship opportunities for U.S. researchers.

CONCLUSIONS

The UNOLS consortium management structure is sound and is of benefit to research institutions, federal agencies, and state and private interests. The federal agency partnerships that capitalize and support the academic research fleet, particularly between the Navy and NSF, successfully provide cost savings and asset sharing. However, some U.S. fleet assets, most notably those operated by NSF's Office of Polar Programs and by NOAA, are not fully integrated with UNOLS. This results in an apparent mismatch between research needs to support national goals and the efficient use of research ship assets, a trend that could continue in the future. NOAA's future ship time needs could potentially be alleviated by increased use of the UNOLS fleet, and opportunity exists to better coordinate the polar fleet supported and operated by OPP and the USCG with the UNOLS fleet.

UNOLS federal agency partners currently participate in international meetings and ship bartering, providing the U.S. research community access to other countries' oceanographic research assets. Increased coordination and integration in the future would strengthen the worldwide community and provide more opportunities to use the UNOLS ship assets efficiently.

7

Conclusions and Recommendations

The U.S. academic research fleet is an essential, enabling resource for the nation. Versatile, capable ships provide the U.S. oceanographic community with access to the sea and the ability to carry out research projects of increasingly critical societal relevance that promote national oceanographic goals. Growth in understanding the ocean's role in climate change, ocean acidification, and marine ecosystem health, among others, will require a robust, technologically capable, and highly adaptable fleet. *Scientific demands on the U.S. academic fleet are likely to increase in future years. However, aging ships and evolving technology require fleet modernization and recapitalization to maintain the nation's leadership in ocean research.* There has been a lack of commitment to previous fleet renewal plans, which has resulted in significant delays to developing a robust academic research fleet.

Recommendation: Federal agencies supporting oceanographic research should implement one comprehensive, long-term research fleet renewal plan to retain access to the sea and maintain the nation's leadership in addressing scientific and societal needs.

The fleet of the future will be required to support increasingly complex, multidisciplinary, multi-investigator research projects, including those in support of autonomous technologies, ocean observing systems, process studies, remote sensing, and modeling. Ship-based research will remain a necessary aspect of oceanographic research in the future (Chapter 2). Although technological advances in remote and autonomous platforms have led to great spatial and temporal increases in sampling, with more gains anticipated, there are still portions of the deep ocean critical to studies of climate change (the water column below 2000 meters, for example) that require sampling by ships. Research vessels are also needed for tracer experiments, for measurement of chemical components of the ocean that do not currently have sensors capable of autonomous use, and for studies of deep sea biodiversity and geology. The largest research vessels of the fleet will be required for global oceanographic surveys. In addition, ship-based calibration and validation will continue to be essential for both over-the-side instruments and satellite remote sensing data streams. Coastal regions that experience the greatest human impacts will need capable Regional and smaller class vessels. Research vessels will also be needed for geological explorations of the seafloor, including large-scale seafloor mapping, seismic surveys, and drilling. Finally, the academic fleet will continue to play a unique and essential role in atmospheric chemistry research programs, providing access to the marine atmosphere with a duration and payload unmatched by other platforms.

New technologies are likely to increase the need for research ships that are capable of supporting multidisciplinary, multi-investigator science (Chapter 3). Research vessels of the future will increasingly be used as platforms that coordinate the operations of multiple autonomous vehicles and/or remotely operated vehicles, deployment of over-the-side instruments, and collection of complex datasets. Highly qualified marine support staff will be increasingly required for successful cruises. *Ocean observatories and autonomous vehicles will impact future vessel design requirements for acoustic communications, deck space, payload, berthing, launch and recovery, and stability.* Precise positioning will be needed to support off-board vehicles. Deployment, recovery, and maintenance of autonomous vehicles, remotely operated vehicles, and moorings that support long-term ocean observatories will require adaptable, technologically capable ships with large laboratory and deck spaces. *Servicing ocean observatories and launching and recovering autonomous vehicles will result in increased demands for ship time. There is a need for increased ship-to-shore bandwidth, in order to facilitate real-time, shore-based modeling and data analysis in support of underway programs, allow more participation of shore-based scientists, and increase opportunities for outreach.*

Oceanographic research needs and advances in technology will drive many aspects of future oceanographic ship design (Chapter 4), increasing laboratory, deck space, and berthing. Research vessel design must accommodate evolving research trends and unforeseen technological advances, while continuing to meet specific disciplinary needs. *Supporting future research needs will require both highly adaptable general purpose ships and spe-*

cialized vessels. The need to investigate societally relevant research questions in remote areas and inclement weather conditions will require some vessels that are capable of operating in high latitudes and high sea states. More capable Coastal, Regional, and Global class ships will also be needed.

Research vessels acquired through the Navy have had little opportunity for scientific community involvement regarding design needs and specifications. *Development of the National Science Foundation (NSF)-sponsored Alaska Regional Research Vessel (ARRV) has benefited from community-driven ship design.*

Recommendation: All future UNOLS ship acquisitions, beginning with the planned Ocean class vessels, should involve the scientific user community from the preconstruction phase through post-delivery of the ship.

The U.S. research fleet has recently faced increasing operating costs and declining days at sea, a trend that is likely to continue (Chapter 5). Primary drivers of operational costs include crew salaries and benefits, fuel and lube oil, and ship scheduling. Ship scheduling will become increasingly efficient to accommodate the needs of the scientific research community. However, tighter schedules for the future fleet could reduce the potential for late-breaking scientific and funding opportunities and increase the wait time for project starts. The trend toward multi-investigator science programs indicates continued need for ship resources. *The increasing cost of ship time and the economies of scale associated with larger ships may lead to greater use of the Global class vessels, which have laboratories, deck space, and berthing capabilities that can support multiple science operations.* This would enable projects to be overlapped and combined into a single leg, thereby driving down the cost per project and per required science berth. To fully realize savings, future ships must be increasingly capable of carrying out multiple science operations simultaneously.

Recommendation: The future academic research fleet requires investment in larger, more capable, general purpose Global and Regional class ships to support multidisciplinary, multi-investigator research and advances in ocean technology.

The University-National Oceanographic Laboratory System (UNOLS) consortium management structure is sound and is of benefit to research institutions, federal agencies, and state and private interests (Chapter 6). *The federal agency partnerships that capitalize and support the academic research fleet, particularly between the Navy and NSF, successfully provide cost savings and asset sharing. However, there are many assets that are not integrated with UNOLS, leading to suboptimal use of the full U.S. research fleet.* Further integration and coordination with agencies that operate and support academic research vessels outside of the UNOLS consortium would optimize use of the entire U.S. research fleet.

Recommendation: NOAA should identify which of its 13,200 unmet annual ship day needs could be supported by the UNOLS fleet. NOAA and UNOLS should work together to develop a long-term plan to increase the usage of UNOLS ships in support of the NOAA mission.

Recommendation: The NSF Division of Ocean Sciences, the NSF Office of Polar Programs, and the U.S. Coast Guard should improve coordination of ship operations and support between the UNOLS and polar research fleets.

References

Baker, P. and M. McNutt. 1996. FUMAGES: The Future of Marine Geology and Geophysics: Report of a Workshop. National Science Foundation, Ashland Hills, Oregon.

Bank, M.S., E. Chesney, J.P. Shine, A. Maage, and D.B. Senn. 2007. Mercury Bioaccumulations and Trophic Transfer in Sympatric Snapper Species from the Gulf of Mexico. *Ecological Applications* 17:2100-2110.

Bash, J.F. 2001. University-National Oceanographic Laboratory System. *Oceanography* 14(2):23-25.

Betzer, P., D. Hayes, R. Knox, C. Mooers, R. Pittenger, and R. Wall. 2005. Projections for UNOLS' Future - Substantial Financial Challenges. UNOLS.

Bigelow, H.B. 1931. *Oceanography: Its Scope, Problems, and Economic Importance*. The Riverside Press, Cambridge, Massachusetts, 236pp.

Bole, M. and C. Forrest. 2005. Early Stage Integrated Parametric Ship Design. *Proceedings ICCAS 2005*, 447-460.

Bowen, A.D., D.R. Yoerger, C. Taylor, R. McCabe, J. Howland, D. Gomez-Ibanez, J.C. Kinsey, M. Heintz, G. McDonald, D.B. Peters, B. Fletcher, C. Young, J. Buescher, L.L. Whitcomb, S.C. Martine, S.D. Webster, and M.V. Jakuba. 2008. The Nereus Hybrid Underwater Robotic Vehicle for Global Ocean Science Operations to 11,000m Depth. *Proceedings of IEEE/MTS Oceans 2008*, Quebec City, Canada.

Bucklin, A. 2000. Methods for Population Genetic Analysis of Zooplankton, *The ICES Zooplankton Methodology Manual*. Academic Press, London, United Kingdom, 684pp.

Buffle, J. and G. Horvai (eds.). 2000. *In Situ Monitoring of Aquatic Systems: Chemical Analysis and Speciation*. John Wiley and Sons, New York, 642pp.

Byrne, J.V. and R.P. Dinsmore. 2000. Two Years of Turbulence Leading to a Quarter Century of Cooperation: The Birth of UNOLS. *50 Years of Ocean Discovery: NSF 1950-2000*. National Academy Press, Washington, D.C., 107-116.

Chave, A.D., A. Bowen, S. Glenn, W. Hill, M. Kosro, E. Massion, L. Mayer, D. Schwartz, K. Smith, B. Wall, F.B. Wooding, and P.F. Worcester. 2003. Ocean Observatories Initiative Facilities Needs from UNOLS: Report of the UNOLS Working Group on Ocean Observatory Facility Needs. UNOLS.

Chen, C.Y., N. Serrell, D.C. Evers, B.J. Fleishman, K.F. Lambert, J. Weiss, R.P. Mason, and M.S. Bank. 2008. Meeting Report: Methylmercury in Marine Ecosystems—From Sources to Seafood Consumers. *Environmental Health Perspectives* 116:1706-1712.

Collins, C., W. Gardner, M. McNutt, and P. Ortner. 2006. Criteria and Process for Recommending Non-operational Periods of Ships in the UNOLS Fleet: Report of an Ad-Hoc UNOLS Subcommittee.

Commission on Marine Science, Engineering, and Resources. 1969. Our Nation and the Sea: A Plan for National Actions. U.S. Government Printing Office, Washington, D.C.

Cowles, T. and L. Atkinson. 2000. Assessment of Future Science Needs in the Context of the Academic Oceanographic Fleet. National Science Foundation, Corvallis, Oregon.

Cullen, V. 2009. WHOI: A Brief History. Woods Hole Oceanographic Institution, Woods Hole, Massachusetts.

Curry, W., J. Broda, L. Keigwin, G. Mountain, and N. Pisias. 2008. A New Long Coring System for R/V Knorr. *Eos* 89(15):142-143.

D'Asaro, E.A. 2003. Performance of Autonomous Lagrangian Floats. *Journal of Atmopsheric and Oceanic Technology* 20(6):896-911.

Daidola, J.C. 2004. Oceanographic Research Ships. In Lamb, T. (ed.), *Ship Design and Construction, Volume 2*, Jersey City, New Jersey.

Daly, K., R. Jahnke, M. Moline, R. Detrick, D. Luther, G. Matsumoto, L. Mayer, and K. Raybould. 2006. ORION: Report of the Design and Implementation Workshop. Salt Lake City, Utah.

Davis, R.E. 1991. Observing the General Circulation with Floats. *Deep-Sea Research* 38: 531-571.

Davis, R.E., C.E. Eriksen, and C.P. Jones. 2003. Autonomous Buoyancy Driven Underwater Gliders. In Griffiths, G. (ed.), *Technology and Applications of Autonomous Underwater Vehicles*. Taylor & Francis, London, United Kingdom, 368pp.

Dinsmore, R.P. 1998. History of UNOLS. *Oceanography* 14(2):2001.

European Space Agency. 2009. New Mission Will Provide Global Maps of Soil Moisture and Ocean Salinity. *ScienceDaily*, Available online at http://www.sciencedaily.com/releases/2009/05/090528092732.htm.

Federal Oceanographic Facilities Committee. 2001. Charting the Future for the National Academic Research Fleet: A Long Range Plan for Renewal.

Fisichella, D. 2009. The Future of Science Support at Sea: The Role of the Marine Technician. Power Point presentation, Woods Hole, Massachusetts.

Fleet Review Committee. 1999. The Academic Research Fleet: A Report to the Assistant Director for Geosciences. National Science Foundation, Arlington, Virginia.

Garzoli, S.L., O. Boebel, H. Bryden, R.A. Fine, M. Fukasawa, S. Gladyshev, G. Johnson, M. Johnson, A. MacDonald, C. Meinen, H. Mercier, A. Orsi, A. Piola, S. Rintoul, S. Speich, M. Visbeck, and R. Wanninkhof. 2009. Progressing Towards Global Sustained Deep Ocean Observations. *Ocean Obs 2009*, Venice, Italy.

GEOTRACES Planning Group. 2006. GEOTRACES Science Plan. Scientific Committee on Oceanic Research, Baltimore, Maryland. Available online at http://www.geotraces.org/sciencePlan/documents/GEOTRACESFinalWebVersion.pdf.

Glenn, S. and T. Dickey (eds.). 2003. *Scientific Cabled Observatories for Time Series (SCOTS)*. National Science Foundation Report, Consortium for Oceanographic Research and Education, Washington, D.C., 80pp.

Herr, F. 2006. Value of Partnering via UNOLS for ONR and Navy: Prepared for NRAC Panel on Ocean Class Ship. Power Point presentation, Washington, D.C.

Hood, M., M. Fukasawa, N. Gruber, G.C. Johnson, C. Sabine, B. Sloyan, K. Stansfield, and T. Tanhua. 2009. Ship-Based Repeat Hydrography: A Strategy for a Sustained Global Program. *Ocean Obs 2009*, Venice, Italy.

Hooker, S. 2009. Remote Sensing and Satellite Needs: Being Prepared for the Next Generation NASA Mission. Power Point presentation, Woods Hole, Massachusetts.

Interagency Working Group on Facilities. 2007. Federal Oceanographic Fleet Status Report.

International Maritime Organization. 1997. Information Resources on Air Pollution from Ships' Emissions and the Impact of Climate Change on Shipping. London, United Kingdom.

Jahnke, R., L. Atkinson, J. Barth, F. Chavez, K. Daly, J. Edson, P. Franks, J. O'Donnell, and O. Schofield. 2002. Coastal Ocean Processes and Observatories: Advancing Coastal Research. *Report on the CoOP Observatory Science Workshop*, Savannah, Georgia.

Jahnke, R., J. Bane, A. Barnard, J. Barth, F. Chavez, H. Dam, E. Dever, P. DiGiacomo, J. Edson, R. Geyer, S. Glenn, K. Johnson, M. Moline, J. O'Donnell, J. Oltman-Shay, O. Persson, O. Schofield, H. Sosik, and E. Terrill. 2003. Coastal Observatory Research Arrays: A Framework for Implementation Planning. Report on the CoOP CORA Workshop. Skidaway Institute of Oceanography Technical Report Draft TR-03-01. 69 pp.

Joint Subcommittee on Ocean Science and Technology. 2007. Charting the Course for Ocean Science in the United States for the Next Decade: An Ocean Research Priorities Plan and Implementation Strategy.

Jumars, P. and M. Hay. 1999. Ocean Ecology: Understanding and Vision for Research. *Report of the OEUVRE Workshop*. National Science Foundation, Arlington, Virginia.

Knauss, J.A. 1990. The Stratton Commission - Its History and Its Legacy. *Oceanography* 3(1):53-55.

Lambin, E.F., X. Baulies, N. Bockstael, G. Fischer, T. Krug, R. Leemans, E.F. Moran, R.R. Rindfuss, Y. Sato, D. Skole, B.L. Turner, and C. Vogel. 1999. Land-Use and Land-Cover Change (LUCC) Implementation Strategy. IGBP Report No. 48. Nunes, C. and J.I. Auge (eds.). Stockholm, Sweden.

Liss, P., P. Boyd, E. Cortijo, K. Denman, B. Huebert, T. Jickells, T. Johannessen, G. Komen, D. Kumar, P. Matrai, W. Miller, U. Platt, K. Richardson, P. Schlosser, M. Uematsu, L. Wainer, and D. Wallace. 2004. Science Plan and Implementation Strategy. IGBP Report No. 50. Stockholm, Sweden.

Liu, K.K., L. Atkinson, R. Quinones, and L. Talaue-McManus (eds.). 2009. *Carbon and Nutrient Fluxes in Continental Margins*. Springer, Berlin, Germany, 500 pp.

Malakoff, D. 2008. Grim Forecast for a Fading Fleet. *Science* 307:338-340.

MARGINS Office. 2004. NSF MARGINS Program Science Plans 2004. Lamont-Doherty Earth Observatory, Columbia University, New York.

McGillicuddy, D. 2009. Biological–Physical Interaction and Vessel Use. Power Point presentation, Woods Hole, Massachusetts.

McNutt, M., D. Wiesenburg, and E. Hofmann. 2005. UNOLS Ad-Hoc Committee to Address the Impact of Budget Reductions on Fleet Operations. UNOLS.

MESH Workshop Steering Committee. 2005. Paleoclimate Research Workshop Report. National Science Foundation, Arlington, Virginia. Available online at http://www.mesh.usc.edu/files/Paleoclimate%20Research%202006.pdf.

Mitson, R.B. 1995. Underwater Noise of Research Vessels: Review and Recommendations. *Cooperative Research Report No. 209*. International Council for the Exploration of the Sea, Copenhagen, Denmark.

National Academy of Sciences. 1930. Report of the Committee on Oceanography. Bigelow, H.B. (ed.). Washington, D.C.

National Oceanic and Atmospheric Administration. 2005. 20 Year Research Vision: Understanding Global Ecosystems to Support Informed Decision Making. National Oceanic and Atmospheric Administration, U.S. Department of Commerce, Silver Spring, Maryland.

National Oceanic and Atmospheric Administration. 2008. Research in NOAA: Toward Understanding and Predicting Earth's Environment. A Five Year Plan: Fiscal Years 2008-2012. National Oceanic and Atmospheric Administration, U.S. Department of Commerce, Silver Spring, Maryland.

National Research Council. 1999. *Global Ocean Science: Toward an Integrated Approach*. National Academy Press, Washington, D.C.

National Research Council. 2000a. *Fifty Years of Ocean Discovery: National Science Foundation 1950-2000*. National Academy Press, Washington, D.C.

National Research Council. 2000b. *Illuminating the Hidden Planet: The Future of Ocean Observatory Science*. National Academy Press, Washington, D.C.

National Research Council. 2003a. *Enabling Ocean Research in the 21st Century: Implementation of the Network of Ocean Observatories*. The National Academies Press, Washington, D.C.

National Research Council. 2003b. *Exploration of the Seas: Voyage into the Unknown*. The National Academies Press, Washington, D.C.

National Research Council. 2004. *Future Needs in Deep Submergence Science: Occupied and Unoccupied Vehicles in Basic Ocean Research*. The National Academies Press, Washington, D.C.

National Research Council. 2007. *Polar Icebreakers in a Changing World: An Assessment of U.S. Needs*. The National Academies Press, Washington, D.C.

National Science Foundation. 2001. Ocean Sciences at the New Millennium. National Science Foundation, Arlington, Virginia.

Naval Sea Systems Command. 2009. Ocean Class AGOR System Specification. Washington, D.C.

Navy 225. 2000. Powered by Naval Research, from Discovery to Deployment. *Navy 225*. Belmont Publishing, Clearwater, Florida.

Oceanic Mantle Dynamics Workshop. 2002. Oceanic Mantle Dynamics Implementation Plan: Report of a Community Workshop. Snowbird, Utah. Available online at http://www.whoi.edu/science/GG/omd/pdf/OMD_Workshop_Report.pdf.

Office of Marine and Aviation Operations. 2008. NOAA Ship Recapitalization Plan FY 2010 to FY 2024. National Oceanic and Atmospheric Administration, U.S. Department of Commerce, Silver Spring, Maryland.

Office of Naval Research. 2006. Report to Congress: Requirements and Plans for University National Oceanographic Laboratory System (UNOLS) Fleet Renewal. Office of Naval Research, Arlington, Virginia.

Purdy, G.M., D. Karl, R. Detrick, R. Hyndman, J. Delaney, B. Bekins, K. Denman, N. Mantua, T. Michales, J. Barry, A. Gargett, P. Donaghay, J. Bellingham, and P. Worcester. 2003. RECONN: Regional Cabled Observatory Networks (of Networks). *Report of the Cabled Regional Observatory Workshop*, San Francisco, California.

Ridge 2000 Program. 2001. Ridge 2000 Science Plan. Pennsylvania State University, University Park, Pennsylvania, 48pp.

Rudnick, D.L., R.E. Davis, C.C. Eriksen, D.M. Fratantoni, and M.J. Perry. 2004. Underwater Gliders for Ocean Research. *Journal of the Marine Technology Society* 38:48-59.

Sabine, C.L., R.A. Feely, N. Gruber, R.M. Key, K. Lee, J.L. Bullister, R. Wanninkhof, C. S. Wong, D.W.R. Wallace, B. Tilbrook, F.J. Millero, T.-H. Peng, A. Kozyr, T. Ono, and A.F. Rios. 2004. The Oceanic Sink for Anthropogenic CO2. *Science* 305:367-371.

Sager, W., H. Dick, P. Fryer, and H.P. Johnson. 2003. Report from a Workshop: Requirements for Robotic Underwater Drills in U.S. Marine Geologic Research. Texas A&M University, College Station, Texas.

Schmitt, R.W., J.R. Ledwell, E.T. Montgomery, K.L. Polzin, and J.M. Toole. 2005. Enhanced Diapycnal Mixing by Salt Fingers in the Thermocline of the Tropical Atlantic. *Science* 308(5722):685-688.

Schofield, O. and M.K. Tivey. 2004. ORION: Ocean Research Interactive Observatory Networks. San Juan, Puerto Rico.

Scholin, C., G. Doucette, S. Jensen, B. Roman, D. Pargett, R. Marin, C. Preston, W. Jones, J. Feldman, C. Everlove, A. Harris, N. Alvarado, E. Massion, J. Birch, D. Greenfield, R. Vrijenhoek, C. Mikulski, and K. Jones. 2009. Remote Detection of Marine Microbes, Small Invertebrates, Harmful Algae, and Biotoxins Using the Environmental Sample Processer (ESP). *Oceanography* 22(2):158-167.

Stommel, H. 1989. The Slocum Mission. *Oceanography* 2(1):22-25.

Tangirala, S. and J. Dzielski. 2007. A Variable Buoyancy Control System for a Large AUV. *IEEE Journal of Oceanic Engineering* 32(4):762-771.

Treadwell, T.K., D.S. Gorsline, and R. West. 1988. History of the U.S. Academic Oceanographic Research Fleet and the Sources of Research Ships. UNOLS Fleet Improvement Committee, Texas A&M, College Station, Texas.

Treadwell, T.K., D.S. Gorsline, and R. West. 1989. History of the U.S. Academic Oceanographic Research Fleet and the Sources of Research Ships. *Oceanography* 2(2):36-39.

Trenberth, K. and A. Clarke. 1998. CLIVAR Initial Implementation Plan. *WCRP-103*, Hamburg, Germany.

U.S. Commission on Ocean Policy. 2004. An Ocean Blueprint for the 21st Century Final Report. Washington, D.C.

UNOLS Fleet Improvement Committee. 2003a. Regional Class Science Mission Requirements. University-National Oceanographic Laboratory System, Narragansett, Rhode Island. Available online at http://www.unols.org/committees/fic/smr/regional/rcsmr_version1.html.

UNOLS Fleet Improvement Committee. 2003b. Ocean Class Science Mission Requirements. University-National Oceanographic Laboratory System, Narragansett, Rhode Island. Available online at http://www.unols.org/committees/fic/smr/ocean/ocsmr_version1.html.

UNOLS Fleet Improvement Committee. 2009. UNOLS Fleet Improvement Plan 2009. The UNOLS Academic Research Fleet: Continued Access to the Sea. University-National Oceanographic Laboratory System, Narragansett, Rhode Island. Available online at http://www.unols.org/committees/fic/FIP05/Fleet_Improvement_Plan_2009_Final.pdf.

Varney, M.S. 2000. *Chemical Sensors in Oceanography*. Gordon and Breach Sciences Publishers, Singapore, 333pp.

Willey, D.A., R.A. Fine, R.E. Sonnerup, J.L. Bullister, W. M. Smethie, Jr., and M.J. Warner. 2004. Global Oceanic Chlorofluorocarbon Inventory. *Geophysical Research Letters* 31(L01303).

Williams, R. and M. Hawkins. 2009. The Evolution of Research Vessel Design. Power Point presentation, Washington, D.C.

Wilson, C. 2003. Late Summer Chlorophyll Blooms in the Oligotropic North Pacific Subtropical Gyre. *Geophysical Research Letters* 30(18):1942.

Winokur, B. 2009. Navy UAS: N84 UAS Concept Development METOC Applications. Power Point presentation.

Yoerger, D.R., A.M. Bradley, B.B. Walden, H. Singh, and R. Bachmayer. 1998. Surveying a Subsea Lava Flow Using the Autonomous Benthic Explorer (ABE). *International Journal of Systems Science* 29:1031-1044.

Young, W., J. Barth, K. Brink, W. Dewar, C. Freidrichs, R.X. Huang, M. Kawase, S. Lozier, T. Maxworthy, S. Meacham, J. Moum, E. Ralph, B. Qiu, T. Royer, P. Schlosser, and L. Thompson. 1997. The Future of Physical Oceanography. *Report of the APROPOS Workshop*. National Science Foundation, Monterey, California.

Appendixes

A

The History of the U.S. Academic Research Fleet

Ships have historically been important tools for oceanographic research. The use of ships to observe ocean phenomena dates back centuries to endeavors by Charles Darwin, Captain James Cook, and Lieutenant John Wilkes (Navy 225, 2000). The use of vessels for nongovernmental research in the United States dates to the early 1930s (Treadwell et al., 1989). As awareness of the importance of understanding the ocean grew, so did the need to have access to the sea. This led to acquisition and use of dedicated ships and smaller craft, and eventually to the formation of the University-National Oceanographic Laboratory System (UNOLS) consortium in 1971. Over the years, the composition and management of these research vessels have evolved. This topic is examined in periods from pre-World War II to the present (inspired by Treadwell et al., 1988, 1989).

PRE-WORLD WAR II

The oceanographic research fleet began with a few converted small craft. A 1929 report by the National Academy of Sciences Committee on Oceanography led to the establishment of the Woods Hole Oceanographic Institution (WHOI) and the expansion and strengthening of the Scripps Institution of Oceanography and the University of Washington (National Academy of Sciences, 1930; Bigelow, 1931; Cullen, 2009). The first purpose-built research vessel, the *Atlantis*, was commissioned for WHOI in 1930 (Cullen, 2009).

WORLD WAR II TO 1959

Oceanographic research greatly expanded from a few converted, donated vessels. Research was conducted from a wide range of vessels from military combatants to smaller donated ships. After the war, many of those ships stayed on, and surplus Navy and Army craft were acquired and converted. This period also saw additional institutions added to the research community (notably Texas A&M University, University of Miami, Oregon State University, University of Rhode Island, University of Hawaii, and Lamont Doherty Geological Observatory). The Office of Naval Research (ONR) was the main funding source at the beginning of the period, but gradually gave way to the newly created National Science Foundation (NSF), which was heavily influenced by ONR practices and people (Navy 225, 2000; National Research Council, 2000a).

1960 TO 1980

During this period, many new, purpose-built research vessels were acquired. Older, converted vessels were phased out as newly constructed ships entered the fleet. Federal funding accounted for more than 80 percent of these new vessels. The new ships tended to be larger than those being replaced, a trend that has continued through the present day. There was a huge upsurge of ocean research with major national and international programs, such as the International Geophysical Year, the International Decade of Ocean Expedition, the International Indian Ocean Expedition and the Global Atmospheric Research Program (Treadwell et al., 1988; Dinsmore, 1998; National Research Council, 1999; Byrne and Dinsmore, 2000; Navy 225, 2000). Ships acquired during this time included the Navy-funded vessels *Conrad, Washington*, and *Thompson* in the early to mid-1960s; the Navy-funded Global class ships *Melville* and *Knorr* in the late 1960s and early 1970s; and the NSF-funded Intermediate class vessels *Oceanus, Endeavor*, and *Wecoma* in the mid-1970s. The deep submersible *Alvin* was also designed, built, and tested and incorporated into the academic fleet, affording researchers access to the deep ocean.

Fleet growth came at a price. Ship operating funds did not keep pace with costs (Treadwell et al., 1988; Dinsmore, 1998; Byrne and Dinsmore, 2000). This issue was exacerbated by the gradual reduction of ONR ship-based research and Navy funding. Additionally, ship funding and ship scheduling mechanisms became contentious issues. Until the early 1970s, ships were funded through block grants to the operator institutions, putting researchers at non-operator institutions at a disadvantage. Operator institutions controlled the schedules and science parties, creating a "have and have-not" situation among ocean scientists that was resolved through the creation of the UNOLS consortium (Knauss, 1990; Dinsmore, 1998; Bash, 2001).

1980 TO THE PRESENT

The upsurge of oceanographic research and the construction of new research vessels in the 1960s and 1970s led to a robust and capable academic fleet, but many of these ships were approaching obsolescence by the 1980s. A changing focus toward science research priorities, as well as continuing concern about pacing the Soviet submarine threat, led to the need to again replace aging ships with newer, more capable vessels.

At this time, oceanography was moving toward multidisciplinary, global-scale research projects (Treadwell et al., 1988; Fleet Review Committee, 1999). These large projects (including the World Ocean Circulation Experiment [WOCE], Joint Global Ocean Flux Study [JGOFS], Ridge Interdisciplinary Global Experiments [RIDGE], Global Ocean Ecosystem Dynamics [GLOBEC], and MARGINS) envisioned a need for research vessels capable of conducting long cruises (up to 60 days at sea) with multidisciplinary science parties of 30 or more scientists and technicians. None of the existing UNOLS ships was deemed suitable to meet these demanding criteria.

To meet this need, the Navy built three new Global class vessels in the 1990s (*Thompson*, *Revelle*, and *Atlantis*) and completed major midlife conversions on *Melville* and *Knorr*. *Alvin* was modernized, and a new type of deep ocean tool, the remotely operated vehicle (ROV), was adapted to meet science research requirements. During this time, NSF also funded midlife overhauls of its ships (*Oceanus*, *Endeavor*, *Wecoma*, *Point Sur*, and *Cape Hatteras*). UNOLS fleet modernization continued with the building of the first Ocean class vessel, *Kilo Moana*. It is currently the only SWATH (Small Waterplane Area Twin Hull) vessel in the fleet. Regional and Regional/Coastal class vessels (*Atlantic Explorer*, *Walton Smith*, and *Hugh R. Sharp*) were also brought into the fleet. Most recently, NSF acquired the seismic vessel *Marcus Langseth* to replace the aging *Maurice Ewing*.

Presently, three Intermediate class vessels (*Oceanus*, *Endeavor*, *Wecoma*) are slated to retire in 2010 and two Regional class ships (*Point Sur* and *Cape Hatteras*) come to the end of their projected service lives in 2011 (UNOLS Fleet Improvement Committee, 2009).

B

UNOLS Member Institutions[1]

Alabama Marine Environmental Sciences Consortium
University of Alaska, Fairbanks
Bermuda Institute of Ocean Sciences
Bigelow Laboratory for Ocean Sciences
University of California, San Diego, Scripps Institution of Oceanography
University of California, Santa Barbara
University of California, Santa Cruz
Cape Fear Community College
Caribbean Marine Research Center/Perry Institute for Marine Science
Columbia University, Lamont-Doherty Earth Observatory
University of Connecticut
University of Delaware
Duke University/University of North Carolina
Florida Institute of Oceanography
Florida Institute of Technology
Florida State University
Harbor Branch Oceanographic Institution
Harvard University
University of Hawaii
Hobart and William Smith Colleges
Humboldt State University Marine Laboratory

[1] Operator institutions in bold.

The Johns Hopkins University
School of the Coast and Environment, Louisiana State University
Louisiana Universities Marine Consortium
University of Maine
The Marine Science Consortium
University of Maryland
Massachusetts Institute of Technology
University of Miami, Rosenstiel School of Marine & Atmospheric Sciences
University of Michigan
University of Minnesota, Duluth
Monterey Bay Aquarium Research Institute
Moss Landing Marine Laboratories
Naval Postgraduate School (National Oceanographic Aircraft Facility Operator)
University of New Hampshire
State University of New York at Stony Brook
University of North Carolina at Wilmington
Nova University
Old Dominion University
Oregon State University
University of Puerto Rico
Romberg Tiburon Center for Environmental Studies, San Francisco State University
University of Rhode Island
Rutgers University
San Diego State University
Sea Education Association
Smithsonian Tropical Research Institute
University of South Carolina
University of South Florida
University of Southern California
Southern California Marine Institute
University of Southern Mississippi
University System of Georgia, Skidaway Institute of Oceanography
University of Texas
Texas A&M University
Virginia Institute of Marine Science
University of Washington
University of Wisconsin at Madison
University of Wisconsin at Milwaukee, Great Lakes Water Institute
University of Wisconsin at Superior
Woods Hole Oceanographic Institution

C

Acronyms

ADA	Americans with Disability Act
ADCP	Acoustic Doppler Current Profiler
ARRV	Alaska Region Research Vessel
AUV	Autonomous underwater vehicle
CLIVAR	Climate Variability and Predictability
CTD	Conductivity-temperature-depth sensors
DOE	Department of Energy
DP	Dynamic positioning
EPA	Environmental Protection Agency
FOFC	Federal Oceanographic Facilities Committee
GEO	Geosciences Directorate (NSF)
GLOBEC	Global Ocean Ecosystem Dynamics
GPS	Global Positioning System
HAB	Harmful algal blooms
ICES	International Council for the Exploration of the Sea
IMO	International Maritime Organization
INMARTECH	International Marine Technicians Workshops

IODP	Integrated Ocean Drilling Program
IOOS	Integrated Ocean Observing System
ISOM	International Research Ship Operators' Meeting
IWGF	Interagency Working Group on Facilities
JGOFS	Joint Global Ocean Flux Study
JSOST	Joint Subcommittee on Ocean Science and Technology
MARGINS	MARGINS Program (NSF)
MARPOL	International Convention for the Prevention of Pollution from Ships
MGG	Marine geology and geophysics
MMS	Minerals Management Service
NASA	National Aeronautics and Space Administration
NOAA	National Oceanic and Atmospheric Administration
NOPP	National Oceanographic Partnership Program
NRAC	Naval Research Advisory Committee
NRC	National Research Council
NSF	National Science Foundation
OCE	Ocean Sciences Division (NSF)
OFEG	Ocean Facilities Exchange Group
ONR	Office of Naval Research
OOI	Ocean Observatories Initiative
OPP	Office of Polar Programs (NSF)
OTS	Over the side
POC	Particulate organic carbon
ROV	Remotely operated vehicle
SMR	Science Mission Requirements
SOLAS	Surface Ocean Lower Atmosphere Study
SOSUS	Sound Surveillance System
UAF	University of Alaska, Fairbanks
UAV	Unmanned aerial vehicle
UNOLS	University-National Oceanographic Laboratory System
USACE	U.S. Army Corps of Engineers
USBL	Ultrashort baseline

USCOP	U.S. Commission on Ocean Policy
USCG	U.S. Coast Guard
USGS	U.S. Geological Survey
WHOI	Woods Hole Oceanographic Institution
WOCE	World Ocean Circulation Experiment

D

Committee and Staff Biographies

COMMITTEE

Ronald (Ron) Kiss *(co-chair)* is president emeritus of Webb Institute, a private four-year college providing B.S. degrees in naval architecture and marine engineering. Prior to joining Webb Institute, he was vice president of SYNTEK, assisting the U.S. Navy on the Joint Navy/Defense Advanced Research Projects Agency arsenal ship program and the Navy's aircraft carrier and surface combatant programs. He served as deputy assistant secretary of the Navy for ship programs in the Office of the Assistant Secretary of the Navy (Research, Development and Acquisition) and as executive director of the Amphibious, Auxiliary, Mine and Sealift Directorate at Naval Sea Systems Command. Mr. Kiss spent nearly 20 years with the Maritime Administration, culminating as acting associate administrator for Shipbuilding and Ship Operations. Mr. Kiss is a former member of the Marine Board and served on the National Research Council (NRC) Committee on the Assessment of U.S. Coast Guard Polar Icebreaker Roles and Future Needs. He holds a B.S. degree in naval architecture and marine engineering from Webb Institute and an M.S. in naval architecture from the University of California-Berkeley. He has participated in a number of postgraduate programs at institutions including Harvard University and the Massachusetts Institute of Technology.

Richard (Dick) Pittenger *(co-chair)* has spent his career in naval and research oceanography. During his naval career, he served as Oceanographer of the Navy, director of the Antisubmarine Warfare Program, and commander of Destroyer class warships. Upon retirement, Rear Admiral Pittenger came to Woods Hole Oceanographic Institution (WHOI), where he led the Marine Operations Division. While at WHOI, he oversaw the conversion of the University-National Oceanographic Laboratory System (UNOLS) vessels *Knorr* and *Oceanus,* the addition of R/V *Atlantis*, and the retirement of *Atlantis II.* RADM Pittenger has also worked closely with deep submergence vehicles, including the award of a grant to build a replacement for the *Alvin* manned submersible. RADM Pittenger served on the UNOLS Council from 1992 to 1998 and has been a member and vice-chair of NRC committees on naval research and acoustics. He earned his M.S. in physics, specializing in underwater acoustics, at the Naval Postgraduate School.

Francisco Chavez is a senior researcher in the Biological Oceanography Group at the Monterey Bay Aquarium Research Institute (MBARI). His current research focuses on biology and chemistry of the ocean in relation to global change; how climate, ocean physics, marine chemistry, and ocean ecosystems co-vary on global to mesoscales; instrumentation and systems for long-term ocean observing; and satellite remote sensing. His current projects include studies in the equatorial Pacific, central California, and Peru. Dr. Chavez is the associate editor of *Geophysical Research Letters*, and is a member of the National Science Foundation (NSF) Geosciences Advisory Committee and the U.S. Joint Global Ocean Flux Study (JGOFS) time series oversight committee. Dr. Chavez received his Ph.D. in botany from Duke University in 1987.

Margo Edwards is a senior research scientist and director of the Hawaii Mapping Research Group with the Hawaii Institute of Geophysics and Planetology at the University of Hawaii at Manoa. Her current scientific research focuses on bathymetric and sidescan sonar mapping of the Arctic Basin and the use of high-resolution photographic and acoustic data to map the East Pacific Rise mid-ocean ridge. Dr. Edwards was recently appointed to the Scientific Ice Expedition Science Advisory Committee, a collaborative project between the U.S. Navy and civilian scientists for geological and environmental research in the Arctic Ocean. She served as chair of the UNOLS Arctic Icebreaker Coordinating Committee from 2004 to 2007 and on the NRC Committee on Designing an Arctic Observing Network. Dr. Edwards earned her Ph.D. in marine geology and geophysics from Columbia University in 1992.

Rana Fine is a professor of marine and atmospheric chemistry at the University of Miami's Rosenstiel School of Marine and Atmospheric Science. Her current research objective is to better understand the role of the oceans in climate change, occurring on time scales of up to decades. She is interested in the physical processes that determine the oceans' capacity to take up atmospheric constituents such as carbon dioxide, especially through air-sea interactions and ocean mixing. She was the elected president of the Ocean Sciences Section of the American Geophysical Union from 1996-1998, and served on the World Ocean Circulation Experiment (WOCE) Scientific Steering Committee. Dr. Fine is a former member of the Ocean Studies Board and has served on several NRC committees related to oceanography. She received her Ph.D. from the University of Miami in 1975.

Nancy Rabalais is executive director and professor at the Louisiana Universities Marine Consortium. Dr. Rabalais' research includes the dynamics of hypoxic environments, interactions of large rivers with the coastal ocean, estuarine and coastal eutrophication, and environmental effects of habitat alterations and contaminants. Dr. Rabalais is a fellow of the American Association for the Advancement of Science (AAAS), an Aldo Leopold Leadership Program fellow, a national associate of the National Academies of Science, a past president of the Estuarine Research Federation, a vice chair of the Scientific Steering Committee of Land-Ocean Interactions in the Coastal Zone/International Geosphere-Biosphere Programme, a past chair of the NRC Ocean Studies Board, and a current member of the UNOLS Council. She received the 2002 Ketchum Award for coastal research from the Woods Hole Oceanographic Institution and shares the Blasker award with R.E. Turner. She was awarded the American Society of Limnology and Oceanography Ruth Patrick Award and the National Water Research Institute Clarke Prize in summer 2008. Dr. Rabalais received her Ph.D. in zoology from the University of Texas at Austin in 1983.

Eric Saltzman is a professor in the Earth System Science School of Physical Sciences at the University of California, Irvine. Dr. Saltzman's research interests are in atmospheric chemistry, biogeochemistry, and air-sea exchange. His research examines how biologically produced gases in the surface ocean have a major impact on global atmospheric cycling of elements such as sulfur, nitrogen, and carbon and can play an important role in the global climate system. This research involves the development of analytical instruments for trace gas measurement, collection of field data using ships and aircraft, and use of computer models to estimate rates of air-sea exchange and atmospheric reactions. Dr. Saltzman obtained his Ph.D. from the Rosenstiel School of Marine and Atmospheric Science at the University of Miami in 1986.

James Swift is a research oceanographer and academic administrator at the University of California, San Diego Scripps Institution of Oceanography (SIO). Dr. Swift has been on 28 blue water and icebreaker expeditions in the Atlantic, Pacific, Arctic, and Southern Oceans. His primary scientific interests are Arctic water masses and circulation, the global thermohaline circulation, and ocean measurement and interpretation. Dr. Swift is scientific adviser to the SIO Oceanographic Data Facility and coordinator for academic institutions involved in the U.S. Global Ocean Carbon and Repeat Hydrography program. He is also director of the international Climate Variability and Predictability program (CLIVAR) and Carbon Hydrographic Data Office. Dr. Swift was the founding chair of the UNOLS Arctic Icebreaker Coordinating Committee, which oversees science-related aspects of the construction and testing of the research icebreaker U.S. Coast Guard *Healy*, and whose long-term mission includes promoting a productive and successful working relationship between the Coast Guard and the science community using icebreakers. He served as the committee chair from 1996 to 2000. He served on the U.S. Antarctic Research Vessel Oversight Committee and is the former chair of the NSF Office of Polar Programs Advisory Committee. He received his Ph.D. in physical oceanography from the University of Washington.

William Wilcock is a professor of marine geophysics in the School of Oceanography at the University of Washington. His research focuses on the use of seismic techniques to understand submarine volcanoes and mid-ocean ridge hydrothermal systems. Dr. Wilcock's current projects include the installation of a seafloor seismometer network and seismic tomography of the Endeavour Segment, Juan de Fuca Ridge and Deception Island, Antarctic Peninsula. He is actively involved in the development of the regional component of the Ocean Observatories Initiative, a cabled underwater research facility located in the Northeast Pacific Ocean. Dr. Wilcock has an interest in the development of seafloor cabled observatories and is actively involved in research related to the NEPTUNE Project. Dr. Wilcock received his Ph.D. in marine geology and geophysics in 1992 from the Massachusetts Institute of Technology (MIT)/WHOI Joint Program.

Dana Yoerger is a senior scientist in the Applied Ocean Physics and Engineering Department at the Woods Hole Oceanographic Institution. His research focuses on robotics, with a specialization in the design and operation of remotely operated and autonomous vehicles (ROVs and AUVs). His current work involves precise control, navigation, and positioning of vehicles and applying principles of automation to add capability and ease of use. Dr. Yoerger has worked on several ROV and AUV systems, including the *Jason/Medea* ROV, the *Autonomous Benthic Explorer* AUV, the *Sentry* AUV, and development of the hybrid ROV-AUV *Nereus*. He has been to sea on more than 50 oceanographic expeditions, including the 1985 *Titanic* discovery cruise. Dr. Yoerger obtained his Ph.D. in mechanical engineering from MIT in 1982.

STAFF

Deborah Glickson is an associate program officer with the Ocean Studies Board. She received an M.S. in geology from Vanderbilt University in 1999 and a Ph.D. in oceanography from the University of Washington in 2007. Her doctoral research focused on magmatic and tectonic contributions to mid-ocean ridge evolution and hydrothermal activity at the Endeavour Segment of the Juan de Fuca Ridge. Dr. Glickson was a Dean John A. Knauss Marine Policy Fellow and worked on coastal and ocean policy and legislation in the U.S. Senate. Prior to her Ph.D. work, she was a research associate in physical oceanography at Woods Hole Oceanographic Institution. She joined the National Academies staff in 2008 and has worked on *Oceanography in 2025: Proceedings of a Workshop* in addition to this report.

Jeremy Justice is a program assistant with the Ocean Studies Board. Mr. Justice graduated from the University of Oklahoma in 2008 with a B.A. in international and area studies. He joined the National Academies in October 2008.